Leadership in Energy and Environmental Design

LEED® BD&C Practice Exam

Building Design & Construction

Meghan Peot, MEd and Brennan Schumacher, LEED AP

Professional Publications, Inc. • Belmont, California

Benefit by Registering this Book with PPI

- Get book updates and corrections
- Hear the latest exam news
- Obtain exclusive exam tips and strategies
- Receive special discounts

Register your book at **www.ppi2pass.com/register**.

Report Errors and View Corrections for this Book

PPI is grateful to every reader who notifies us of a possible error. Your feedback allows us to improve the quality and accuracy of our products. You can report errata and view corrections at **www.ppi2pass.com/errata**.

LEED® and USGBC® are registered trademarks of the U.S. Green Building Council. PPI® is not affiliated with the U.S. Green Building Council (USGBC) or the Green Building Certification Institute (GBCI) and does not administer the LEED AP program or LEED Green Building Rating System. PPI does not claim any endorsement or recommendation of its products or services by USGBC or GBCI.

Energy Star® (ENERGY STAR®) is a registered trademark of the U.S. Environmental Protection Agency (EPA).

LEED BD&C PRACTICE EXAM: BUILDING DESIGN & CONSTRUCTION

Current printing of this edition: 1

Printing History

edition number	printing number	update
1	1	New book.

Copyright © 2009 by Professional Publications, Inc. (PPI). All rights reserved. No part of this publication may be reproduced, stored in a retrieval system, or transmitted, in any form or by any means, electronic, mechanical, photocopying, recording, or otherwise, without the prior written permission of the publisher.

Printed in the United States of America

PPI
1250 Fifth Avenue, Belmont, CA 94002
(650) 593-9119
www.ppi2pass.com

ISBN: 978-1-59126-185-8

Table of Contents

Preface and Acknowledgments v

Introduction vii
    About the LEED Credentialing Program vii
    About the LEED AP Building Design & Construction Exam vii
    Taking the LEED Credentialing Exams ix
    How to Use This Book x

References xi
    Primary References for Exam Part 1: LEED Green Associate xi
    Secondary Reference for Exam Part 1: LEED Green Associate xi
    References for Exam Part 2: LEED AP Building Design & Construction xii

Practice Exam Part One 1

Practice Exam Part Two 23

Practice Exam Part One Solutions 49

Practice Exam Part Two Solutions 71

Preface and Acknowledgments

The LEED exam changes implemented in 2009 were the catalyst for us to overhaul the questions in our best-selling *LEED NC Sample Exam* and create this book, *LEED BD&C Practice Exam*. Those overhauled questions, plus more than 120 newly developed and incorporated questions and answers, make this the best simulation of the LEED AP BD&C credentialing exam available. The best way to prepare for the exam is to answer exam-like questions such as the ones published here. Though these questions are similar to those on the actual exam, these are original to us and based on situations and questions we have encountered while consulting and instructing.

Brennan would like to thank his partner David Nelson, of David Nelson & Associates, LLC, for the flexibility, time, and support he provided throughout this project. Meghan would like to recognize Jordy Oleson and Bridget Peot for their support in the process of writing this book. Together we wish to acknowledge those at PPI who have helped create this book, including editorial director Sarah Hubbard, production director Cathy Schrott, typesetter and cover designer Amy Schwertman, project editor Courtnee Crystal, and proofreader Megan Synnestvedt.

Finally, your suggestions are important to us. As the LEED exam process evolves, so will this book. If your experiences lead you to different answers than those presented in this book, we'd like to hear from you. Someday, there will be a second edition of this book and your comments will help shape it. We request that you report all errata and suggestions using the errata section on the PPI website at **www.ppi2pass.com/errata**. Corrections will be posted on the PPI website and incorporated into this book when it is reprinted.

Meghan Peot, MEd and Brennan Schumacher, LEED AP

Introduction

About the LEED Credentialing Program

The Green Building Certification Institute (GBCI) offers credentialing opportunities to professionals who demonstrate knowledge of Leadership in Energy and Environmental Design (LEED) green building practices. *LEED BD&C Practice Exam: Building Design & Construction* prepares you for both parts of the LEED AP Building Design & Construction exam.

GBCI's LEED credentialing program has three tiers. The first tier corresponds to the LEED Green Associate exam. According to the *LEED Green Associate Candidate Handbook*, this exam confirms that you have the knowledge and skills necessary to understand and support green design, construction, and operations. When you pass the LEED Green Associate exam, you will earn the LEED Green Associate credential.

The second tier, which corresponds to the LEED AP specialty exams, confirms your deeper and more specialized knowledge of green building practices. GBCI currently has planned five tracks for the LEED AP exams: LEED AP Homes, LEED AP Operations & Maintenance, LEED AP Building Design & Construction, LEED AP Interior Design & Construction, and LEED AP Neighborhood Development. The LEED AP exams are based on the corresponding LEED reference guide and rating systems and other references. When you pass the LEED Green Associate exam along with any LEED AP specialty exam, you will earn the LEED AP credential.

The third tier, called LEED AP Fellow, will distinguish professionals with an exceptional depth of knowledge, experience, and accomplishments with LEED green building practices. This distinction will be attainable through extensive LEED project experience, not by taking an exam.

For more information about LEED credentialing, visit **www.ppi2pass.com/LEEDhome**.

About the LEED AP Building Design & Construction Exam

The LEED AP Building Design & Construction exam contains 200 questions (so does the practice exam in this book). Exam Part 1: LEED Green Associate Exam contains 100 questions that test your knowledge of green building practices and principles, as well as your familiarity with LEED requirements, resources, and processes related to both commercial and residential spaces and both new construction and existing building projects. Accordingly, GBCI categorizes the Exam Part 1 questions into the following seven subject areas.

- *Synergistic Opportunities and LEED Application Process* (project requirements; costs; green resources; standards that support LEED credit; credit interactions; Credit Interpretation Requests and rulings that lead to exemplary performance credits; components of LEED Online and project registration; components of LEED Scorecard; components of letter templates; strategies to achieve credit; project boundary; LEED boundary; property boundary; prerequisites and/or minimum program requirements for LEED certification; preliminary rating; multiple certifications for same building; occupancy requirements; USGBC policies; requirements to earn LEED AP credit)
- *Project Site Factors* (community connectivity: transportation and pedestrian access; zoning requirements; development: heat islands)
- *Water Management* (types and quality of water; water management)
- *Project Systems and Energy Impacts* (environmental concerns; green power)
- *Acquisition, Installation, and Management of Project Materials* (recycled materials; regionally harvested and manufactured materials; construction waste management)
- *Stakeholder Involvement in Innovation* (integrated project team criteria; durability planning and management; innovative and regional design)
- *Project Surroundings and Public Outreach* (codes)

Exam Part 2: LEED AP BD&C Specialty Exam contains an additional 100 questions that test your knowledge of subject areas unique to green building design and construction. Accordingly, GBCI categorizes the Exam Part 2 questions into the following seven subject areas.

- *Project Site Factors* (considerations for site selection: land issues, plants, and animals; community connectivity and services; development: building and land, lighting; climate conditions)
- *Water Management* (water treatment; stormwater; irrigation demand)
- *Project Systems and Energy Impacts* (energy performance policies; building components; on-site renewable energy; requirements for third-party relationships; energy performance measurement; energy trade-offs; sources; energy usage)
- *Acquisition, Installation, and Management of Project Materials* (building reuse; rapidly renewable materials; acquisition of materials)
- *Improvements to the Indoor Environment* (minimum ventilation requirement; tobacco smoke control; air quality; ventilation effectiveness; indoor air quality: pre-construction, during construction, before occupancy, and during occupancy; low-emitting materials; indoor/outdoor chemical and pollutant control; lighting controls; thermal controls; views; types of building spaces)
- *Stakeholder Involvement in Innovation* (design workshop/charrette; ways to earn credit; education of building manager)
- *Project Surroundings and Public Outreach* (infrastructure; zoning requirements; government planning agencies; reduced parking methods; transit-oriented development; pedestrian oriented streetscape design)

Introduction

Taking the LEED Credentialing Exams

To apply for a LEED credentialing exam, you must agree to the disciplinary policy and credential maintenance requirements and submit to an application audit. To be eligible to take the LEED Green Associate exam, one of the following must be true.

- Your line of work is in a sustainable field.
- You have documented experience supporting a LEED-registered project.
- You have attended an education program that addresses green building principles.

To be eligible to take a LEED AP exam, you must have documented experience with a LEED-registered project within the three years prior to your application submittal.

The LEED credentialing exams are administered by computer at Prometric test sites. Prometric is a third-party testing agency with over 250 testing locations in the United States and hundreds of centers globally. To schedule an exam, you must first apply at www.gbci.org to receive an eligibility ID number. Then, you must go to the Prometric website at www.prometric.com/gbci to schedule and pay for the exam. If you need to reschedule or cancel your exam, you must do so directly through Prometric.

The LEED credentialing exam questions are multiple choice with four or more answer options for each question. If more than one option must be selected to correctly answer a question, the question stem will indicate how many options you must choose. Each 100-question exam lasts two hours, giving you a bit more than one and a half minutes per question. The bulk of the questions are non-numerical. Calculators are not allowed or provided, but only basic math is needed to correctly solve any quantitative questions. No reference materials or other supplies may be brought into the exam room, though a pencil and scratch paper will be provided by the testing center. (References are not provided.) The only thing you need to bring with you on exam day is your identification.

Your testing experience begins with an optional brief tutorial to introduce you to the testing computer's functions. When you've finished the tutorial, questions and answer options are shown on a computer screen, and the computer keeps track of which options you choose. Because points are not deducted for incorrectly answered questions, you should mark an answer to every question. For answers you are unsure of, make your best guess and flag the question for later review. If you decide on a different answer later, you can change it, but if you run out of time before getting to all your flagged questions, you still will have given a response to each one. Be sure to mark the correct number of options for each question. There is no partial credit for incomplete answers (or for selecting only some of the correct options).

If you are taking both the first tier (LEED Green Associate) and the second tier (LEED AP) exams on the same day, at the end of your first session the computer will ask you if you are ready to take the second tier. You can take a short break at this time. The second tier's two hours will begin when you click "yes" to indicate that you are ready.

To ensure that the chances of passing remain constant regardless of the difficulty of the specific questions administered on any given exam, GBCI converts the raw exam score to a scaled score, with the total number of points set at 200 and a minimum passing score of 170. In this way, you are not penalized if the exam taken is more difficult than another exam. Instead, in such a case, fewer questions must be answered correctly to achieve a passing score. Your scaled score (or scores, if you are taking both tiers on the same day) is reported on the screen upon completing the exam. A brief optional exit survey completes the exam experience.

When you pass the LEED Green Associate exam, a LEED Green Associate certificate will be sent to you in the mail. If you take and pass both exams, a LEED AP certificate will be sent to you in the mail. If you take both exams but pass only the LEED AP exam, you will need to register again and retake and pass the LEED Green Associate exam before you receive any LEED credential.

How to Use This Book

There are two ways you can use this book's practice exam. You can determine your areas for further study with an untimed review of the questions and answers. Familiarize yourself with the exam format and content and determine which subjects you are weak in. This book's companion volume, *LEED Prep BD&C: What You Really Need to Know to Pass the LEED AP Building Design & Construction Exam*, will give you a complete, concise review of the subjects covered on the exam. *LEED BD&C Flashcards*, also available through PPI, will reinforce your ability to retain and recall what you've studied.

Or, you can use this book to simulate the exam experience, either as a pretest before you begin your study or when you think you are fully prepared. In this case, treat this practice exam as though it were the real thing. Don't look at the questions or answers ahead of time. Put away your study materials and references, set a timer for two hours, and answer as many questions as you can within the time limit. Practice exam-like time management. Fill in the provided bubble sheet with your best guess on every question regardless of your certainty and mark the answers to revisit if time permits. If you finish before the time is up, review your work. If you are unable to finish within the time limit, make a note of where you were after two hours, but continue on to complete the exam. Keep track of your time to see how much faster you will need to work to finish the actual exam within two hours.

After taking the practice exam, check your answers against the answer key. Consider a question correctly answered only if you have selected all of the required options (and no others). Calculate the percentage correct. Though the actual exam score will be scaled, aim for getting at least 70% (70 questions) of the practice exam's questions correct. The fully explained answers are a learning tool. Therefore, in addition to reading the answers to the questions you answered incorrectly, also read the explanations of those you answered correctly. Categorize your incorrect answers by exam subject to help you determine which areas need further study. Use GBCI's list of references and PPI's *LEED Prep BD&C* to guide your preparation. Though this practice exam reflects the breadth and depth of the content on the actual exam, use your best judgment when determining the subjects you need to review.

References

LEED BD&C Practice Exam: Building Design & Construction is based on the following references, identified by the Green Building Certification Institute (GBCI) in its *LEED AP Building Design & Construction Candidate Handbook*. Most of these references are available electronically. You can find links to these references on PPI's website, **www.ppi2pass.com/ LEEDreferences**.

Primary References for Exam Part 1: LEED Green Associate

Bernheim, Anthony, and William Reed. "Part II: Pre-Design Issues." *Sustainable Building Technical Manual*. Public Technology, Inc. 1996.

Cost of Green Revisited: Reexamining the Feasibility and Cost Impact of Sustainable Design in Light of Increased Market Adoption. Sacramento, CA: Davis Langdon, 2007.

Guidance on Innovation & Design (ID) Credits. Announcement. Washington, DC: U.S. Green Building Council, 2004.

Guidelines for CIR Customers. Announcement. Washington, DC: U.S. Green Building Council, 2007.

LEED for Building Design & Construction Reference Guide. Washington, DC: U.S. Green Building Council, 2009.

LEED Technical and Scientific Advisory Committee. *The Treatment by LEED of the Environmental Impact of HVAC Refrigerants*. Washington, DC: U.S. Green Building Council, 2004.

Secondary References for Exam Part 1: LEED Green Associate

AIA Integrated Project Delivery: A Guide. American Institute of Architects, 2007.

Americans with Disabilities Act: Standards for Accessible Design. 28 CFR Part 36. Washington, DC: Code of Federal Regulations, 1994.

"Codes and Standards." Washington, DC: International Code Council, 2009.

"Construction and Building Inspectors." *Occupational Outlook Handbook.* Washington, DC: Bureau of Labor Statistics, 2009.

Frankel, Mark, and Cathy Turner. *Energy Performance of LEED for New Construction Buildings: Final Report.* Vancouver, WA: New Buildings Institute and U.S. Green Building Council, 2008.

GSA 2003 Facilities Standards. Washington, DC: General Services Administration, 2003.

Guide to Purchasing Green Power: Renewable Electricity, Renewable Energy Certifications, and On-Site Renewable Generation. Washington, DC: Environmental Protection Agency, 2004.

Kareis, Brian. *Review of ANSI/ASHRAE Standard 62.1-2004: Ventilation for Acceptable Indoor Air Quality.* Greensboro, NC: Workplace Group, 2007.

Lee, Kun-Mo, and Haruo Uehara. *Best Practices of ISO 14021: Self-Declared Environmental Claims.* Suwon, Korea: Ajou University, 2003.

LEED Steering Committee. *Foundations of the Leadership in Energy and Environmental Design Environmental Rating System: A Tool for Market Transformation.* Washington, DC: U.S. Green Building Council, 2006.

References for Exam Part 2: LEED AP Building Design & Construction Specialty

Bernheim, Anthony, and William Reed. "Part II: Pre-Design Issues." *Sustainable Building Technical Manual.* Public Technology, Inc. 1996.

Cost of Green Revisited: Reexamining the Feasibility and Cost Impact of Sustainable Design in Light of Increased Market Adoption. Sacramento, CA: Davis Langdon, 2007.

Frankel, Mark, and Cathy Turner. *Energy Performance of LEED for New Construction Buildings: Final Report.* Vancouver, WA: New Buildings Institute and U.S. Green Building Council, 2008.

Guidance on Innovation & Design (ID) Credits. Announcement. Washington, DC: United States Green Building Council, 2004.

Guidelines for CIR Customers. Announcement. Washington, DC: U.S. Green Building Council, 2007.

LEED Online Sample Credit Templates. Washington, DC: U.S. Green Building Council, 2009.

LEED for Building Design & Construction Reference Guide. Washington, DC: U.S. Green Building Council, 2009.

Practice Exam Part One

Exam time limit: 2 hours

LEED BD&C Practice Exam

1. Ⓐ Ⓑ Ⓒ Ⓓ	35. Ⓐ Ⓑ Ⓒ Ⓓ	68. Ⓐ Ⓑ Ⓒ Ⓓ
2. Ⓐ Ⓑ Ⓒ Ⓓ	36. Ⓐ Ⓑ Ⓒ Ⓓ	69. Ⓐ Ⓑ Ⓒ Ⓓ
3. Ⓐ Ⓑ Ⓒ Ⓓ	37. Ⓐ Ⓑ Ⓒ Ⓓ	70. Ⓐ Ⓑ Ⓒ Ⓓ
4. Ⓐ Ⓑ Ⓒ Ⓓ	38. Ⓐ Ⓑ Ⓒ Ⓓ	71. Ⓐ Ⓑ Ⓒ Ⓓ
5. Ⓐ Ⓑ Ⓒ Ⓓ	39. Ⓐ Ⓑ Ⓒ Ⓓ	72. Ⓐ Ⓑ Ⓒ Ⓓ
6. Ⓐ Ⓑ Ⓒ Ⓓ	40. Ⓐ Ⓑ Ⓒ Ⓓ	73. Ⓐ Ⓑ Ⓒ Ⓓ
7. Ⓐ Ⓑ Ⓒ Ⓓ	41. Ⓐ Ⓑ Ⓒ Ⓓ	74. Ⓐ Ⓑ Ⓒ Ⓓ
8. Ⓐ Ⓑ Ⓒ Ⓓ	42. Ⓐ Ⓑ Ⓒ Ⓓ	75. Ⓐ Ⓑ Ⓒ Ⓓ
9. Ⓐ Ⓑ Ⓒ Ⓓ	43. Ⓐ Ⓑ Ⓒ Ⓓ	76. Ⓐ Ⓑ Ⓒ Ⓓ Ⓔ Ⓕ
10. Ⓐ Ⓑ Ⓒ Ⓓ	44. Ⓐ Ⓑ Ⓒ Ⓓ	77. Ⓐ Ⓑ Ⓒ Ⓓ Ⓔ Ⓕ
11. Ⓐ Ⓑ Ⓒ Ⓓ	45. Ⓐ Ⓑ Ⓒ Ⓓ Ⓔ	78. Ⓐ Ⓑ Ⓒ Ⓓ
12. Ⓐ Ⓑ Ⓒ Ⓓ	46. Ⓐ Ⓑ Ⓒ Ⓓ Ⓔ Ⓕ	79. Ⓐ Ⓑ Ⓒ Ⓓ
13. Ⓐ Ⓑ Ⓒ Ⓓ Ⓔ	47. Ⓐ Ⓑ Ⓒ Ⓓ Ⓔ	80. Ⓐ Ⓑ Ⓒ Ⓓ
14. Ⓐ Ⓑ Ⓒ Ⓓ	48. Ⓐ Ⓑ Ⓒ Ⓓ	81. Ⓐ Ⓑ Ⓒ Ⓓ
15. Ⓐ Ⓑ Ⓒ Ⓓ	49. Ⓐ Ⓑ Ⓒ Ⓓ	82. Ⓐ Ⓑ Ⓒ Ⓓ
16. Ⓐ Ⓑ Ⓒ Ⓓ	50. Ⓐ Ⓑ Ⓒ Ⓓ	83. Ⓐ Ⓑ Ⓒ Ⓓ Ⓔ
17. Ⓐ Ⓑ Ⓒ Ⓓ	51. Ⓐ Ⓑ Ⓒ Ⓓ	84. Ⓐ Ⓑ Ⓒ Ⓓ Ⓔ Ⓕ
18. Ⓐ Ⓑ Ⓒ Ⓓ	52. Ⓐ Ⓑ Ⓒ Ⓓ	85. Ⓐ Ⓑ Ⓒ Ⓓ Ⓔ Ⓕ
19. Ⓐ Ⓑ Ⓒ Ⓓ	53. Ⓐ Ⓑ Ⓒ Ⓓ Ⓔ Ⓕ	86. Ⓐ Ⓑ Ⓒ Ⓓ Ⓔ
20. Ⓐ Ⓑ Ⓒ Ⓓ Ⓔ	54. Ⓐ Ⓑ Ⓒ Ⓓ	87. Ⓐ Ⓑ Ⓒ Ⓓ
21. Ⓐ Ⓑ Ⓒ Ⓓ Ⓔ	55. Ⓐ Ⓑ Ⓒ Ⓓ	88. Ⓐ Ⓑ Ⓒ Ⓓ Ⓔ Ⓕ
22. Ⓐ Ⓑ Ⓒ Ⓓ Ⓔ	56. Ⓐ Ⓑ Ⓒ Ⓓ Ⓔ	89. Ⓐ Ⓑ Ⓒ Ⓓ Ⓔ
23. Ⓐ Ⓑ Ⓒ Ⓓ	57. Ⓐ Ⓑ Ⓒ Ⓓ	90. Ⓐ Ⓑ Ⓒ Ⓓ
24. Ⓐ Ⓑ Ⓒ Ⓓ	58. Ⓐ Ⓑ Ⓒ Ⓓ	91. Ⓐ Ⓑ Ⓒ Ⓓ Ⓔ
25. Ⓐ Ⓑ Ⓒ Ⓓ	59. Ⓐ Ⓑ Ⓒ Ⓓ Ⓔ	92. Ⓐ Ⓑ Ⓒ Ⓓ
26. Ⓐ Ⓑ Ⓒ Ⓓ	60. Ⓐ Ⓑ Ⓒ Ⓓ	93. Ⓐ Ⓑ Ⓒ Ⓓ
27. Ⓐ Ⓑ Ⓒ Ⓓ	61. Ⓐ Ⓑ Ⓒ Ⓓ	94. Ⓐ Ⓑ Ⓒ Ⓓ
28. Ⓐ Ⓑ Ⓒ Ⓓ Ⓔ	62. Ⓐ Ⓑ Ⓒ Ⓓ	95. Ⓐ Ⓑ Ⓒ Ⓓ Ⓔ
29. Ⓐ Ⓑ Ⓒ Ⓓ	63. Ⓐ Ⓑ Ⓒ Ⓓ	96. Ⓐ Ⓑ Ⓒ Ⓓ Ⓔ Ⓕ
30. Ⓐ Ⓑ Ⓒ Ⓓ Ⓔ	64. Ⓐ Ⓑ Ⓒ Ⓓ	97. Ⓐ Ⓑ Ⓒ Ⓓ
31. Ⓐ Ⓑ Ⓒ Ⓓ	65. Ⓐ Ⓑ Ⓒ Ⓓ	98. Ⓐ Ⓑ Ⓒ Ⓓ
32. Ⓐ Ⓑ Ⓒ Ⓓ	66. Ⓐ Ⓑ Ⓒ Ⓓ Ⓔ	99. Ⓐ Ⓑ Ⓒ Ⓓ
33. Ⓐ Ⓑ Ⓒ Ⓓ	67. Ⓐ Ⓑ Ⓒ Ⓓ	100. Ⓐ Ⓑ Ⓒ Ⓓ Ⓔ
34. Ⓐ Ⓑ Ⓒ Ⓓ		

Practice Exam Part One

1. As part of the LEED documentation process, a mechanical engineer is responsible for which of the following?

    (A) declaring compliance with either NPDES or local control standards
    (B) demonstrating potable water use reductions by water-consuming fixtures
    (C) demonstrating that stormwater management strategies reduce stormwater runoff
    (D) listing structural controls with pollutant removal methods and percent annual rainfall treated

2. Underground petroleum storage tanks are discovered during the design phase of a warehouse to be constructed near a wetland. The owners would like to know if the site is a brownfield and if they are eligible for grants to fund the site's cleanup. Which of the following CANNOT classify a LEED site as a brownfield?

    (A) local governmental agency
    (B) civil engineer
    (C) county court system
    (D) U.S. EPA national or regional office

3. Which of the following organizations provides building industry acoustical standards?

    (A) Acoustical and Board Products Association
    (B) Air Diffusion Council
    (C) American National Standards Institute
    (D) American Institute of Architects

4. Project teams may NOT pursue multiple LEED certifications under which of the following circumstances?

    (A) without written authorization from USGBC
    (B) simultaneously
    (C) unless one of the certifications is under the LEED EBO&M rating system
    (D) unless the building is registered with The National Register of Historic Places

5. ANSI/ASHRAE 62.1 sets standards for which of the following?

    (A) greenhouse gases
    (B) heating
    (C) lighting
    (D) ventilation

6. Which of the following is NOT necessary to complete the baseline water calculation for potable use?

    (A) building floor area
    (B) water volume per fixture use
    (C) number of full-time equivalent male occupants
    (D) types of water-using fixtures

7. Emissivity is the ratio of which of the following?

    (A) radiation emitted by a surface to the radiation emitted by a black body
    (B) interior illuminance at a given plane to exterior illuminance
    (C) reflected solar energy to incoming solar energy
    (D) transmitted light to the total incident of light

8. To which of the following projects would the LEED for Commercial Interiors rating system BEST apply?

    (A) construction of a 9000 sq ft (840 sq meter) commercial real estate office building on a previously developed site
    (B) retrofit of the mechanical and electrical systems of an existing commercial tenant space
    (C) renovation of a tenant space from a hair salon into a coffee shop
    (D) construction of an office complex for future commercial tenants

9. Using flyash in concrete will contribute most directly toward credits that encourage which of the following?

    (A) using recycled-content building materials
    (B) using low-VOC building materials
    (C) optimizing energy performance
    (D) using salvageable building materials

10. Which of the following is NOT a factor used in determining LEED certification costs?

    (A) USGBC membership
    (B) total area of the project
    (C) date of submission for review
    (D) LEED rating system used

11. Some types of refrigerants used in HVAC & R equipment are considered ozone depleting compounds (ODCs). To reduce ODCs, production of which of the following refrigerants was halted in 1995 in the United States?

    (A) HCFCs
    (B) HFCs
    (C) CFCs
    (D) Halons

12. Which of the following best matches USGBC's description of a *tradeoff*?
    - (A) a strategy for achieving multiple credits by pursuing credits them separately rather than together
    - (B) a credit-earning strategy that requires the sacrifice of a particular LEED prerequisite
    - (C) a situation in which pursuing multiple strategies requires a compromise
    - (D) a credit-earning strategy that involves the exchange of building materials for multiple LEED projects

13. A project's mechanical engineer is typically responsible for providing documentation for which of the following? (Choose two.)
    - (A) stormwater design
    - (B) outdoor air delivery monitoring systems
    - (C) low-emitting materials specifications
    - (D) construction waste management
    - (E) refrigerant management

14. Which of the following will BEST help reduce water consumption?
    - (A) automated toilet sensors
    - (B) dual-flush toilets
    - (C) pressure assisted toilets
    - (D) blow-out fixtures

15. Which of the following is the first step toward LEED certification?
    - (A) meet with a GBCI representative
    - (B) complete the online registration
    - (C) find a LEED AP to sponsor the project
    - (D) submit the precertification application

16. At what point in the LEED documentation process is it necessary to submit cut sheets containing volatile organic compound (VOC) data?
    - (A) when the commissioning agent requests it
    - (B) when OSHA requests it
    - (C) when GBCI performs an audit
    - (D) when the LEED letter templates are submitted

17. Which of the following vehicles is NOT considered a low-emitting or fuel-efficient vehicle within the context of LEED?

    (A) gas-electric hybrid car with a green score of 30 from the American Council for an Energy Efficient Economy vehicle rating guide

    (B) hybrid van with a zero emission classification by the California Air Resources Board

    (C) propane-powered truck with green score of 40 from the American Council for an Energy Efficient Economy vehicle rating guide

    (D) SUV with ZEV classification by the California Air Resources Board

18. The Energy Policy Act of 1992 provides standards associated with which of the following?

    (A) stormwater management
    (B) plumbing fixtures
    (C) ozone protection
    (D) energy modeling

19. The LEED BD&C reference guide is best described as which of the following?

    (A) referenced standard guide with abstracts of environmental policies
    (B) sustainable design guide and user's manual for the LEED for New Construction, Core and Shell, and Schools rating systems
    (C) sustainable material resource guide
    (D) green building directory of LEED APs, LEED GAs, and LEED APs with specialty

20. Which of the following are benefits of LEED certification? (Choose two.)

    (A) discounted attendance to all USGBC-sponsored events
    (B) discounted re-certification under the LEED Existing Buildings: Operations and Maintenance rating system
    (C) market exposure through the USGBC website
    (D) online self-certification of green building performance measures
    (E) third-party validation of a building's performance

21. The USGBC website and LEED Online provide many green building resources. Which of the following can be found on these websites? (Choose two.)

    (A) case studies of LEED projects
    (B) directories of green building materials
    (C) searchable list of LEED APs
    (D) samples of green building specifications
    (E) links to all organizations referenced within the LEED reference guides

22. Which of the following does ASHRAE 55 deal with? (Choose two.)

    (A) energy efficiency
    (B) humidity
    (C) indoor air quality
    (D) lighting
    (E) temperature

23. Which of the following does USGBC define as "end-user waste that has become feedstock for another product"?

    (A) post-collection recycled content
    (B) post-consumer recycled content
    (C) post-industrial recycled content
    (D) post-production recycled content

24. A project team is using an integrated design approach to comply with green building standards set forth in the LEED rating systems. For which of the following will points be credited toward a project's LEED certification?

    (A) having all project team members attend a LEED training workshop
    (B) meeting minimum requirements of ANSI/ASHRAE 62.1's ventilation rate procedure and approved addenda
    (C) meeting ANSI/ASHRAE/IESNA 90.1 or a more stringent local energy code
    (D) not using refrigerants

25. A project team does NOT need to document which of the following with an Innovation in Design credit submittal?

    (A) intent of the proposed innovation credit
    (B) proposed requirements of compliance
    (C) ongoing measurement and verification protocol
    (D) strategies used to meet the requirements

26. Project teams must be consistent in the documented number of full-time equivalent (FTE) building occupants for LEED credit submittals. According to USGBC, a full-time worker has an FTE value of how many people?

    (A) 0.5
    (B) 0.8
    (C) 1.0
    (D) 2.0

27. ANSI/ASHRAE/IESNA 90.1, a common building industry standard, sets standards for which of the following?

    (A) measuring energy performance of buildings
    (B) measuring air change effectiveness
    (C) testing ventilation filters for removal efficiency
    (D) measuring ventilation rates

28. Site development, grading, and clearing can cause significant erosion. Which of the following are stabilization control methods? (Choose two.)

    (A) constructing an earth dike
    (B) mulching
    (C) permanent seeding
    (D) installing a silt fence
    (E) installing a sediment trap

29. The minimum efficiency report value is based on which of the following?

    (A) water efficiency
    (B) air filtration
    (C) energy efficiency
    (D) light pollution

30. During the design phase of a LEED project, the owner is responsible for establishing which of the following? (Choose two.)

    (A) designated smoking areas, if smoking is not prohibited
    (B) control zones for the carbon dioxide monitoring system
    (C) location of the recycling area
    (D) project team's intention to meet only the minimum local zoning requirements for parking
    (E) baseline for water use calculations

31. Which of the following is NOT a prerequisite for projects seeking LEED certification?

    (A) reduce CFCs in HVAC & R equipment
    (B) control environmental tobacco smoke
    (C) monitor outdoor air delivery
    (D) reduce water use

32. Who is responsible for assigning LEED Online team member roles for submitting LEED letter templates?

    (A) GBCI staff
    (B) project architect
    (C) owner
    (D) project administrator

33. Which of the following organizations developed a standard that sets limits on VOCs?

    (A) ASHRAE
    (B) IESNA
    (C) SCAQMD
    (D) SMACNA

34. In which of the following LEED credit categories would a point be awarded for establishing staging areas during construction on a greenfield site?

    (A) Sustainable Sites
    (B) Materials and Resources
    (C) Energy and Atmosphere
    (D) Regional Priority

35. A project team is concerned about the amount of heat gain on the project site. The green building consultant has suggested that the team use materials that have a high albedo. Which of the following describes albedo?

    (A) amount of light reflected by a material
    (B) ratio of reflected solar energy to incident solar energy
    (C) radiation emitted by a surface to the radiation emitted by a black body
    (D) solar glazing on a window

36. USGBC offers many resources on its website. One such resource is a collection of LEED case studies. Which of the following is NOT provided in these LEED case studies?

    (A) LEED scorecard
    (B) project photographs
    (C) project statistics
    (D) strategies and results

37. An architect recommends during the programming and schematic phases that the project be registered as soon as possible. Once registered, the project team will receive which of the following?

    (A) email support provided by a TAG
    (B) discounts on certification fees
    (C) access to the credit interpretation rulings database
    (D) two complementary credit interpretation rulings

38. Which of the following materials does NOT have pre-consumer recycled content, according to USGBC?

    (A) composite board made from sawdust from a sawmill
    (B) carpeting made from rubber from a tire plant
    (C) carpeting made from recycled plastic bottles
    (D) feed made from spent grain from a brewery

39. For credits within the Materials and Resources credit category, total materials cost is determined as either the sum of actual material costs based on _____ in CSI Masterformat 2004 Divisions 3-10, 31, and 32; or as a percentage of the total construction costs based on _____ in CSI Masterformat 2004 Divisions 3-10, 31, and 32.

    (A) hard costs only, hard costs only
    (B) hard costs only, total costs
    (C) total costs, hard costs only
    (D) total costs, total costs

40. A project team is working to earn points for reducing water use in the irrigation systems, and is looking for sources of graywater to use for site irrigation. Which of the following is a source of graywater?

    (A) toilets
    (B) food preparation sinks
    (C) bathroom sinks
    (D) waterless urinals

41. SMACNA provides guidelines for protecting HVAC systems and maintaining acceptable indoor air quality during construction. Which of the following project team members would oversee the application of SMACNA guidelines?

    (A) electrical engineer and structural engineer
    (B) commissioning agent and civil engineer
    (C) general contractor and owner
    (D) mechanical engineer and general contractor

42. A LEED project team is considering purchasing a percentage of a building's energy from a renewable energy resource as defined by the Center for Resource Solutions. This is a strategy for meeting the requirements of which of the following LEED credits?

    (A) On-Site Renewable Energy
    (B) Green Power
    (C) Optimize Energy Performance
    (D) Rapidly Renewable Resources

43. According to the Energy Policy Act of 1992, what is the daily water use of 10 full-time equivalent (FTE) female occupants using a conventional water closet?

    (A) 3 uses at 1.2 gallons (4.5 liters)
    (B) 30 uses at 1.6 gallons (6.1 liters)
    (C) 40 uses at 1.2 gallons (4.5 liters)
    (D) 40 uses at 1.6 gallons (6.1 liters)

44. Which of the following statements about a credit interpretation request (CIR) is true?

    (A) Teams must format the CIR as a letter including background information about the project and the questioned credit.
    (B) The rulings on a CIR guarantee a credit award.
    (C) The CIR and ruling must be submitted with the LEED application.
    (D) Teams must include the credit name and team contact information when submitting a CIR.

45. The appeal is an optional step of the LEED certification and documentation process. Decisions regarding which of the following CANNOT be appealed? (Choose two.)

    (A) minimum program requirements
    (B) second level appeals
    (C) credits
    (D) rulings on credit interpretation requests
    (E) prerequisites

46. Exceeding the established credit requirements by a percentage established by a LEED rating system can make a project eligible for point earning in which of the following LEED credit categories? (Choose two.)

    (A) Exemplary Performance
    (B) Innovation in Design
    (C) Innovation in Operations
    (D) Measurement and Verification
    (E) Optimized Credit Achievement
    (F) Regional Priority

47. Which of the following is NOT specified by USGBC as a potential use for process water? (Choose two.)

    (A) boilers
    (B) cooling towers
    (C) irrigation systems
    (D) chillers
    (E) plumbing systems

48. The cost of which of the following can be included in the total cost of construction materials for LEED projects?

    (A) overhead
    (B) elevators
    (C) furniture
    (D) plumbing fixtures

49. Minimum Acoustical Performance is a prerequisite for which LEED BD&C rating system?

    (A) LEED for Core and Shell
    (B) LEED for Schools
    (C) LEED for New Construction
    (D) LEED for Commercial Interiors

50. USGBC defines life cycle cost as which of the following?

    (A) the total cost of an asset over its useful or anticipated life
    (B) a pension plan's present cost compared to its present future benefits
    (C) a product's full environmental cost over its lifetime, from the harvesting of raw material to final disposal
    (D) a project's pre-construction and construction costs

51. Registrations costs for LEED certification vary depending on which of the following?

    (A) USGBC membership
    (B) total area of the building
    (C) building type
    (D) LEED rating system used

52. The full-time equivalent (FTE) occupancy calculation is based on all of the following EXCEPT which?

    (A) building capacity
    (B) number of peak transients
    (C) number of part-time employees
    (D) number of residents

53. Which of the following are minimum project requirements (MPRs)? (Choose three.)

    (A) preventing greenwashing
    (B) streamlining management of the LEED documentation process
    (C) providing a static baseline for potential projects throughout the evolution of the LEED rating systems
    (D) giving clear guidance to customers
    (E) protecting the integrity of the LEED program
    (F) reducing challenges that occur during the LEED certification process

54. LEED regional priority credits are determined based on which of the following?

    (A) area code
    (B) climate zone
    (C) zip code
    (D) annual rainfall

Practice Exam Part One

55. For an individual who has not previously earned a LEED AP credential, which of the following is NOT required to become a LEED AP with specialty?

    (A) passing the LEED Green Associate exam
    (B) participating in a LEED training course
    (C) participating in the credentialing maintenance program
    (D) documenting professional experience on a LEED project within the last three years

56. Which of the following are hard costs associated with construction? (Choose three.)

    (A) architectural design fees
    (B) contractor overhead and profit
    (C) environmental site remediation costs
    (D) labor costs
    (E) materials costs

57. Which of the following lists of construction phases is in correct chronological order?

    (A) schematic design, design development, construction documentation, occupancy certification, final completion
    (B) design development, schematic design, construction documentation, final completion, occupancy certification
    (C) schematic design, design development, construction documentation, final completion, occupancy certification
    (D) pre-design, schematic design, post design, construction documentation, occupancy certification

58. Which of the following tasks CANNOT be completed using LEED Online?

    (A) registering a project
    (B) managing project details
    (C) completing documentation requirements
    (D) submitting applications for review

59. USGBC bases LEED credit weightings on policies by which of the following organizations? (Choose two.)

    (A) U.S. EPA
    (B) NIST
    (C) BREEAM
    (D) ANSI
    (E) AESI

60. ANSI/ASHRAE 52.2 provides methods for which of the following?

    (A) improving the energy performance of buildings
    (B) measuring air change effectiveness
    (C) testing ventilation filters for removal efficiency
    (D) measuring ventilation rates

61. The LEED certification process gives project teams the opportunity to appeal denied credits. After which phase may a credit appeal be made?

    (A) the final design review only
    (B) the final construction review only
    (C) either the final design review or final construction review
    (D) either the preliminary design review or preliminary construction review

62. For how many months must a building be continuously occupied to be considered an existing building?

    (A) 6 months
    (B) 12 months
    (C) 18 months
    (D) 24 months

63. Which of the following is a requirement for retaining the LEED Green Associate credential?

    (A) demonstrating professional experience on a LEED project every year
    (B) serving as apprentice to a LEED AP on a LEED project once every 5 years
    (C) retaking and passing the LEED GA exam within 5 years of the initial pass date
    (D) participating in at least 15 hours of continuing education every 2 years

64. Which of the following statements is true about LEED?

    (A) All rating systems have a maximum of 100 points.
    (B) All credits are worth a minimum of 1 point.
    (C) All credits have point values that are positive, whole numbers.
    (D) All rating systems give credits a single, static weight.

65. Which of the following organizations certifies LEED projects?

    (A) USGBC
    (B) GBCI
    (C) Prometric
    (D) ANSI

66. Which of the following types of water can be used in LEED calculations that include potable water? (Choose two.)

    (A) well water
    (B) water from a municipal system
    (C) harvested rainwater
    (D) water collected from showers and hand sinks
    (E) water recovered from building systems

67. What area is used to determine the density radius?

    (A) project site area
    (B) development footprint area
    (C) building floor area
    (D) pervious surface area

68. *The Cost of Building Green Revisited*, a construction cost report by the Davis Landgon group, draws four conclusions about costs of projects seeking LEED certification as compared to projects not seeking LEED certification. Which of the following is NOT a conclusion from this report?

    (A) Significant variations in building costs exist within the same building program category.
    (B) Cost differences between buildings are due to how green a building is.
    (C) LEED projects cost more than conventional construction projects.
    (D) LEED features should always be treated as upgrades or additions to the project scope.

69. The calculations for Materials and Resources credits dealing with construction waste management are based on which of the following?

    (A) percentage of construction waste reduction
    (B) percentage of waste diverted from landfills
    (C) cost of waste management services
    (D) cost of materials used on the project

70. What minimum percentage of building floor area must be occupied by owner or tenant to obtain LEED for New Construction certification?

    (A) 33%
    (B) 40%
    (C) 45%
    (D) 50%

71. Which of the following projects can be certified only under LEED for Schools?

    (A) a building on a prekindergarten school campus
    (B) an academic building on an elementary school campus
    (C) an academic building on a postsecondary campus
    (D) an administrative office on an elementary school campus

72. Under which of the following rating systems can a New Construction project earn additional certification?

    (A) LEED for Commercial Interiors
    (B) LEED for Core and Shell
    (C) LEED for Existing Buildings: Operations and Maintenance
    (D) no other rating system

73. A life cycle analysis is a method for determining which of the following?

    (A) the total cost of an asset cost over its useful or anticipated life
    (B) a pension plan's present cost compared to its present future benefits
    (C) a product's full environmental cost, from harvesting raw material to final disposal
    (D) the amount of energy consumed in the extraction, manufacture, transport, construction, and assembly of building materials

74. Which of the following best matches USGBC's description of a *synergy*?

    (A) an effect of pursuing multiple strategies together that requires a compromise
    (B) a mechanical system capable of producing renewable energy to two or more buildings
    (C) a situation in which pursuing credits together rather than separately has an added benefit
    (D) a thermodynamic quantity equivalent to the capacity of a physical system to do work

75. Increasing natural light in a building will LEAST likely result in a reduction of which of the following?

    (A) building energy load
    (B) occupant productivity
    (C) glare
    (D) contrast

76. Which of the following site treatments are permeable? (Choose three.)

    (A) concrete
    (B) gravel
    (C) asphalt
    (D) open-grid pavers
    (E) planting beds
    (F) grouted pavers

77. Installing which of the following will help reduce blackwater production? (Choose three.)

    (A) automated faucet sensors
    (B) automated toilet sensors
    (C) dual-flush toilets
    (D) waterless urinals
    (E) composting toilets
    (F) reduced-flow aerated faucets

78. Which of the following is the purpose of implementing an erosion and sedimentation control plan?

    (A) promote biodiversity
    (B) restore damaged areas
    (C) reduce pollution from construction activities
    (D) limit the disruption of natural hydrology

79. Precertification is an option for which of the following LEED rating systems?

    (A) LEED for Core and Shell
    (B) LEED for Existing Buildings
    (C) LEED for New Construction
    (D) LEED for Schools

80. Which of the following designations can be used on the business card of an individual who has passed the LEED Green Associate exam?

    (A) LEED AP Green Associate
    (B) LEED Green Associate
    (C) GBCI LEED Green Associate
    (D) LEED Certified

LEED BD&C Practice Exam

81. How many points must a project registered under a LEED 2009 rating system obtain to achieve certification?

    (A) 39 points
    (B) 40 points
    (C) 49 points
    (D) 50 points

82. To register for LEED certification, a project must meet minimum program requirements (MPRs). Which of the following statements is included in the LEED MPR Guidance document?

    (A) LEED for Commercial Interiors projects must have a minimum gross floor area of 250 sq feet (23 sq meters).
    (B) There is no minimum building area to site area ratio.
    (C) Whole project energy and water usage data must be discretionarily distributed.
    (D) The LEED project may be designed to move to a different location in its lifetime.

83. Which of the following is true of the credit interpretation request (CIR) process? (Choose two.)

    (A) A fee is applied for each CIR submitted.
    (B) CIRs must be submitted in letter format.
    (C) Rulings on CIRs are final and guarantee credit achievement.
    (D) The project team must review the CIR database for rulings on similar requests prior to new CIR submittal.
    (E) The project team must submit a separate CIR for each LEED requirement in question.

84. Who may assist the project team in obtaining the home's preliminary rating under the LEED for Homes rating system? (Choose two.)

    (A) Green Home Consultant
    (B) Green Home Designer
    (C) Green Rater
    (D) LEED AP
    (E) LEED for Homes contractor
    (F) LEED for Homes provider

85. What is a LEED credit's weighting based on? (Choose two.)

    (A) air quality optimization
    (B) energy efficiency
    (C) habitat protection
    (D) water use reduction
    (E) greenhouse gas emission reduction
    (F) synergies and tradeoffs

86. Project teams must include which of the following in their narratives for each online submittal template? (Choose two.)

    (A) names of project team members involved
    (B) timeline of strategy implementation
    (C) description of unique project circumstances related to the credit
    (D) strategy cost
    (E) description of related tradeoffs and synergies

87. LEED certification qualifies projects to exhibit the LEED logo plaque after which of the following steps?

    (A) final construction review completion
    (B) LEED application submission
    (C) final certification review acceptance
    (D) final design review completion

88. A project's civil engineer would typically be responsible for documenting which of the following? (Choose two.)

    (A) water use reduction
    (B) use of fuel-efficient vehicles
    (C) habitat protection or restoration
    (D) stormwater design
    (E) roof heat island reduction
    (F) light pollution reduction

89. Which of the following are NOT examples of soft costs associated with construction? (Choose two.)

    (A) legal fees
    (B) equipment costs
    (C) land costs
    (D) architectural design fees
    (E) engineering fees

90. Which of the following is the most likely outcome when a project team's lighting designer reduces the connected load?

    (A) The lighting power density will decrease and the cooling load will decrease.
    (B) The lighting power density will decrease and the cooling load will increase.
    (C) The lighting power density will increase and the cooling load will decrease.
    (D) The lighting power density will increase and the cooling load will increase.

91. The daylighting design on the south facade of a building in a warm climate includes inadequate shading on large vision glazing windows with a high visible light transmittance. These combined factors will most likely result in an increase in which of the following? (Choose three.)

    (A) thermal gain
    (B) glare
    (C) lighting power density
    (D) contrast
    (E) glazing factor

92. Why are geothermal energy systems considered on-site renewable?

    (A) They convert hot water or steam into electrical power.
    (B) They transfer thermal resources from the earth into the building HVAC systems.
    (C) They provide a central thermal energy source to a group of buildings.
    (D) They use vapor compression systems to produce heat for the building's HVAC needs.

93. What type of vegetation would be best to use on a project in a hot and humid region if the project team intends to minimize water use?

    (A) drought-tolerant plants
    (B) hardy native plants
    (C) native plants with moisture sensors
    (D) woody plants and evergreens

94. Passive ventilation is most successfully implemented in which of the following climates?

    (A) arid
    (B) tropical
    (C) temperate
    (D) highland

95. Prohibiting smoking to meet the Environmental Tobacco Smoke prerequisite in the Indoor Environmental Quality credit category can contribute toward achieving which other green building objectives? (Choose two.)

    (A) minimizing climate change
    (B) conserving natural resources
    (C) minimizing energy use
    (D) minimizing occupant exposure to chemical pollutants
    (E) providing a comfortable thermal environment

96. What project data are needed to verify compliance with LEED Development Density and Community Connectivity credits? (Choose two.)

    (A) building floor area
    (B) property area
    (C) number of FTE building occupants
    (D) development footprint area
    (E) parking lot area
    (F) building footprint area

97. A school is being constructed on a site classified as contaminated according to a Phase I Environmental Site Assessment. If the project team still wants to use the site, what is the next step in the process of becoming LEED certified?

    (A) Conduct a Phase II Environmental Site Assessment.
    (B) Provide an Erosion and Sedimentation Plan.
    (C) Remediate the site.
    (D) No further action can be taken as the project cannot become LEED certified.

98. A city has released a request for proposals that includes requiring the project team to satisfy the LEED Accredited Professional (AP) credit in the Innovation and Design credit category. Which of the following must be completed to satisfy the credit's requirements?

    (A) The design firm must employ one LEED AP.
    (B) The project's commissioning agent must be a LEED AP.
    (C) All principal members of the project team must be LEED APs.
    (D) The project team must have at least one LEED AP as a principal member.

99. The LEED BD&C rating systems set minimum erosion and sedimentation control standards as a prerequisite in the Sustainable Sites credit category. Which of the following must the project team comply with to meet these standards?

    (A) 2003 U.S. EPA Construction General Permit
    (B) ASTM E1903-97 Phase II Environmental Site Assessment
    (C) U.S. Code of Federal Regulations, Title 7, Volume 6
    (D) U.S. Code of Federal Regulations, Title 40

100. A LEED project's general contractor is in the best position to provide data and documentation for which of the following? (Choose two.)

    (A) stormwater design
    (B) construction waste management
    (C) energy performance optimization
    (D) storage and collection of recyclables produced by building occupants
    (E) low-emitting materials use

Practice Exam Part Two

Exam time limit: 2 hours

LEED BD&C Practice Exam

1. A B C D
2. A B C D E F
3. A B C D E F
4. A B C D
5. A B C D
6. A B C D E
7. A B C D
8. A B C D E F
9. A B C D E F
10. A B C D
11. A B C D E
12. A B C D
13. A B C D
14. A B C D E
15. A B C D
16. A B C D
17. A B C D
18. A B C D E F
19. A B C D
20. A B C D E
21. A B C D
22. A B C D
23. A B C D
24. A B C D
25. A B C D E
26. A B C D E F
27. A B C D
28. A B C D
29. A B C D E
30. A B C D
31. A B C D E
32. A B C D
33. A B C D
34. A B C D E F
35. A B C D
36. A B C D
37. A B C D
38. A B C D
39. A B C D E
40. A B C D E F
41. A B C D
42. A B C D
43. A B C D
44. A B C D
45. A B C D
46. A B C D
47. A B C D
48. A B C D
49. A B C D
50. A B C D E
51. A B C D E F G
52. A B C D
53. A B C D E F G
54. A B C D E F
55. A B C D E F G
56. A B C D E
57. A B C D
58. A B C D
59. A B C D E
60. A B C D
61. A B C D
62. A B C D E
63. A B C D
64. A B C D
65. A B C D
66. A B C D
67. A B C D
68. A B C D
69. A B C D E F
70. A B C D
71. A B C D
72. A B C D
73. A B C D E F G
74. A B C D E
75. A B C D E F
76. A B C D
77. A B C D
78. A B C D
79. A B C D E
80. A B C D E
81. A B C D
82. A B C D
83. A B C D
84. A B C D E
85. A B C D
86. A B C D
87. A B C D
88. A B C D
89. A B C D
90. A B C D
91. A B C D
92. A B C D
93. A B C D
94. A B C D E
95. A B C D
96. A B C D
97. A B C D E F
98. A B C D E F G
99. A B C D E
100. A B C D

Practice Exam Part Two

1. A midwestern law office building with 15,000 sq ft (1400 sq meters) of perimeter private offices, 30,000 sq ft (2800 sq meters) of non-perimeter open offices, and 10,000 sq ft (900 sq meters) of non-regularly occupied spaces is being designed with occupant comfort as a main priority. To meet specific temperature and humidity conditions for the building, the project team is complying with ASHRAE 55. However, such compliance may prevent the team from also meeting the requirements of which of the following LEED for New Construction credits?

   (A) EA Credit 1, Optimize Energy Performance
   (B) EA Credit 5, Measurement and Verification
   (C) IEQ Credit 4.1, Low-Emitting Materials: Adhesives and Sealants
   (D) IEQ Credit 6.1, Controllability of Systems: Lighting

2. A 20,000 sq ft (1900 sq meter) environmental learning center will be constructed on a 130,500 sq ft (12100 sq meter) site with an existing asphalt parking lot. The remainder of the site will be preserved and dedicated to vegetation and wildlife habitats. The project team will use insulating concrete form (ICF) wall systems and renewable resources such as bamboo and cork. Signage will teach visitors about on-site green building technologies and strategies. These efforts can contribute toward meeting the requirements of which of the following LEED BD&C credits? (Choose four.)

   (A) SS Credit 4.1, Alternative Transportation: Public Transportation Access
   (B) SS Credit 5.2, Site Development: Maximize Open Space
   (C) EA Credit 1, Optimize Energy Performance
   (D) EA Credit 2, On-Site Renewable Energy
   (E) MR Credit 6, Rapidly Renewable Materials
   (F) ID Credit 1, Innovation in Design

3. The project team for a 14,000 sq ft (1300 sq meter) retail shop in an existing building will use black porous pavement systems made from pre-consumer recycled material. This strategy can contribute toward meeting the requirements of which of the following LEED BD&C credits? (Choose two.)

   (A) SS Credit 5, Site Development
   (B) SS Credit 6, Stormwater Design
   (C) SS Credit 7, Heat Island Effect
   (D) SS Credit 8, Light Pollution Reduction
   (E) MR Credit 4, Recycled Content
   (F) MR Credit 5, Regional Materials

4. Plans for a high-performance 80,000 sq ft (7400 sq meter) laboratory include the use of photoelectric daylight sensors. Using the sensors will contribute toward meeting the requirements of which LEED BD&C credit?

   (A) SS Credit 8, Light Pollution Reduction
   (B) EA Credit 1, Optimize Energy Performance
   (C) EA Credit 2, On-Site Renewable Energy
   (D) IEQ Credit 8, Daylight and Views

5. A project team is considering options for compliance with EA Credit 1, Optimize Energy Performance. The project is a new 75,000 sq ft (7000 sq meter) warehouse with an additional 19,500 sq ft (1800 sq meter) office. Photovoltaics are being used as a source of renewable energy. The project team intends to implement enhanced commissioning processes, as well as four energy conservation measures. Which of the following EA Credit 1 options would be most suitable for this project?

    (A) option 1's whole building energy simulation

    (B) option 2's prescriptive compliance path: ASHRAE, Advanced Energy Design Guide

    (C) option 3's prescriptive compliance path: Advanced Buildings™ Core Performance™ Guide

    (D) option 4's prescriptive compliance path: Advanced Buildings Benchmark™ Version 1.1

6. The project team of a 48,000 sq ft (4500 sq meter), mixed-use project in New Mexico intends to implement water conservation strategies. The current design calls for a vegetated roof, composting toilets, low-flow sinks and showers, and a drip irrigation system. Which of the following are needed to determine the reduction in potable water used for site irrigation for WE Credit 1, Water Efficient Landscaping? (Choose two.)

    (A) amount of water treated and conveyed by a public agency specifically for non-potable uses

    (B) vegetated roof area

    (C) existing site imperviousness

    (D) density factor

    (E) species factor

7. The owners of a 19,000 sq ft (1800 sq meter) community recreation center that is under construction want to measure the center's ongoing energy use. The project team is striving for EA Credit 5, Measurement and Verification. Since the project is small, the project team has decided to develop a measurement and verification plan in compliance with the International Performance Measurement and Verification Protocol (IPMVP)'s Option B: Energy Conservation Measure Isolation. To comply with the requirements of EA Credit 5, which of the following should the project team use to establish baseline energy consumption under hypothetical post-construction operating conditions?

    (A) calculations

    (B) energy simulation

    (C) ANSI/ASHRAE/IESNA 90.1

    (D) Department of Energy Buildings Consumption Survey

8. Project teams can earn Innovation in Design points for exemplary performance for which of the following credits? (Choose three.)

    (A) SS Credit 5.2, Site Development: Maximize Open Space
    (B) SS Credit 8, Light Pollution Reduction
    (C) WE Credit 1, Water-Efficient Landscaping
    (D) WE Credit 2, Innovative Wastewater Technologies
    (E) IEQ Credit 8.1, Daylight and Views: Daylight
    (F) IEQ Credit 2, Increased Ventilation

9. A 30,000 sq ft (2800 sq meter) office building in an eco-industrial park is being designed to comply with SS Credit 8, Light Pollution Reduction. Of the following, which information must the team compile to meet the requirements of this credit? (Choose three.)

    (A) area of exterior pavement surfaces
    (B) operation sequence of emergency nighttime lighting
    (C) lamp lumens for exterior luminaires
    (D) location of property line
    (E) watts per square foot (square meter) of exterior illumination
    (F) watts per square foot (square meter) of interior illumination

10. For a project over 50,000 sq ft (4600 sq meters) to meet the requirements of EA Prerequisite 1, Fundamental Commissioning of the Building Energy Systems, which person is the MOST suitable commissioning agent?

    (A) the construction manager
    (B) the mechanical engineer
    (C) a consultant to or employee of the owner
    (D) the mechanical engineer's subcontractor responsible for energy modeling

11. The renovation of a three-story, 78,000 sq ft (7200 sq meter) facility will include a major renovation of the existing mechanical system. If the building owner intends to commission the renovation and meet the requirements of EA Credit 3, Enhanced Commissioning, which of the following tasks must be performed by the commissioning agent? (Choose two.)

    (A) conduct a design review of the owner's project requirements, basis of design, and design documents
    (B) develop a systems manual to illustrate optimal operation of commissioned systems
    (C) develop a plan to account for ongoing building energy consumption
    (D) review building operation 10 months after substantial completion
    (E) verify that the requirements for training personnel and building occupants have been completed

LEED BD&C Practice Exam

12. A project is planned for the northwestern United States near Puget Sound. The owners would like to promote on-site bioretention and are considering a number of different low-impact development strategies. Which of the following credits or prerequisites would bioswales contribute toward?

    (A) SS Prerequisite 1, Construction Activity Pollution Prevention
    (B) SS Credit 6, Stormwater Management
    (C) SS Credit 7, Heat Island Effect
    (D) WE Credit 1, Water Efficient Landscaping

13. A Las Vegas construction project will use an interactive display and educational website to educate the public on water conservation. The project's water usage will be 40% less than the baseline established for the water efficiency prerequisite. Given these strategies only, how many points can the project earn under the LEED for New Construction rating system?

    (A) 0 points
    (B) 2 points
    (C) 4 points
    (D) 5 points

14. Designers of a 30,000 sq ft (2800 sq meter) school in a rural setting seek to earn SS Credit 6.2, Stormwater Design: Quality Control. Incorporating which of the following green building design strategies would help them meet the requirements of this credit? (Choose three.)

    (A) infiltration basin
    (B) solar hot water system
    (C) vegetated roof
    (D) high-albedo concrete
    (E) constructed wetland

15. Owners of a research lab are planning a 150,000 sq ft (14,000 sq meter) facility that will include a Living Machine™. Which of the following does the project team NOT need to document to meet the requirements of WE Credit 2, Innovative Wastewater Technologies?

    (A) flow rate on sprinkler heads
    (B) graywater volumes collected
    (C) number of occupants
    (D) number of workdays

16. EA Credit 2, On-Site Renewable Energy, requires that at least 1% of the building's total energy use be supplied from on-site renewable energy systems. According to the LEED BD&C reference guide, which of the following is a renewable energy source?

    (A) passive solar heat
    (B) daylight
    (C) photovoltaic electricity
    (D) heat from ground source heat pumps

17. Many LEED credits require a basis of comparison (a baseline) for evaluating the actual performance of a proposed building that incoporates a particular green building strategy. For which of the following credits is it necessary to establish a baseline?

    (A) EA Credit 1, Optimize Energy Performance, Option 1
    (B) EA Credit 4, Enhanced Refrigerant Management, Option 1
    (C) IEQ Credit 6.1, Controllability of Systems: Lighting
    (D) IEQ Credit 7, Thermal Comfort

18. A local distribution company is designing a new warehouse that will optimize energy performance. To comply with the requirements of EA Credit 5, Measurement and Verification, which of the following must be continuously monitored? (Choose four.)

    (A) boiler efficiencies
    (B) building-related process energy systems and equipment
    (C) daylight factors throughout the building
    (D) indoor water risers and outdoor irrigation systems
    (E) lighting systems and controls
    (F) stormwater runoff volumes

19. MR Credit 6, Rapidly Renewable Materials, requires that 2.5% of the total value of all building materials and products used in a project be rapidly renewable. USGBC defines rapidly renewable materials as those grown or raised and harvested within a cycle of no more than how many years?

    (A) 7 years
    (B) 10 years
    (C) 15 years
    (D) 25 years

20. The owners of an environmental learning center that is being constructed primarily from local lumber would like to comply with the requirements of MR Credit 7, Certified Wood. To verify compliance with this credit, the project team must know the costs associated with which of the following items? (Choose two.)

    (A) salvaged and refurbished wood-based materials
    (B) rough carpentry
    (C) post-consumer recycled wood fiber portion of any product
    (D) wood doors and frames
    (E) pre-consumer recycled wood fiber portion of any product

LEED BD&C Practice Exam

21. A project team has identified the scope of work needed to comply with the requirements of EA Prerequisite 1, Fundamental Commissioning of the Building Energy Systems, and EA Credit 3, Enhanced Commissioning. Which of the following is NOT an example of an enhanced commissioning task under the LEED BD&C rating systems?

    (A) designating a commissioning authority to lead all commissioning activities
    (B) reviewing the construction documents
    (C) identifying the commissioning team and its responsibilities
    (D) reviewing the contractor submittals for the commissioned systems

22. The area of a previously developed site is 135,000 sq ft (12,500 sq meters), and the area of the building footprint is 35,000 sq ft (3300 sq meters). What area of the existing site must the project team restore with native or adaptive species to earn an exemplary performance point for SS Credit 5.1, Site Development: Protect or Restore Habitat?

    (A) 35,000 sq ft (3300 sq meters)
    (B) 45,000 sq ft (4200 sq meters)
    (C) 55,000 sq ft (5100 sq meters)
    (D) 75,000 sq ft (7000 sq meters)

23. Visible transmittance, a factor contributing to the role of windows in a building's daylighting strategy, can be defined as which of the following?

    (A) ratio of the total heat striking a surface to the heat transmitted
    (B) ratio of the light transmitted to the total light hitting the surface
    (C) ratio of the total energy hitting a surface to the light transmitted
    (D) total light transmitted

24. Which of the following products is LEAST likely to contain rapidly renewable materials?

    (A) bamboo flooring
    (B) FSC-certified wood
    (C) linoleum flooring
    (D) wool carpet

25. The requirements of SS Credit 1, Site Selection, stipulate that projects must NOT be developed on which of the following types of sites? (Choose three.)

    (A) prime farmland as defined by the USDA in the U.S. Code of Federal Regulations
    (B) contaminated land or land classified as a brownfield by a local, state, or federal agency
    (C) habitat for any species on federal or state threatened or endangered lists
    (D) previously undeveloped and ungraded land in a natural state, designated as a greenfield
    (E) previously undeveloped land less than 5 feet (1.5 meters) above FEMA's 100-year flood elevation

26. IEQ Credit 6, Controllability of Systems encourages occupant productivity, comfort, and well-being by providing which of the following? (Choose two.)

    (A) additional outdoor air ventilation to improve indoor air quality
    (B) a high level of system control over thermal comfort
    (C) ongoing assessment of occupants' thermal comfort
    (D) ventilation system monitoring
    (E) lighting control
    (F) pollutant source control

27. A project team has decided to pursue design strategies to meet the requirements of MR Credit 1, Building Reuse. Which of the following is included in calculations for this credit?

    (A) structural elements in square feet (square meters)
    (B) shell elements in cubic feet (cubic meters)
    (C) exterior paving materials in cubic feet (cubic meters)
    (D) building footprint in square feet (square meters)

28. A project team working for a university located in a warm climate is interested in reducing the energy consumption of a new student union, which is currently in the design phase. Which of the following strategies would contribute the MOST to accomplishing this goal?

    (A) decreasing the albedo of building hardscape materials
    (B) increasing the lighting power density
    (C) decreasing the solar heat gain coefficient
    (D) increasing the building's ventilation rates

29. Which of the following adhesives or sealants must meet the requirements of IEQ Credit 4.1, Low Emitting Materials: Adhesives and Sealants? (Choose two.)

    (A) roofing adhesives
    (B) fire stopping sealants
    (C) stucco adhesives
    (D) aerosol adhesives
    (E) exterior concrete sealants

30. A project is using $100,000 worth of new wood and $50,000 worth of reclaimed wood for flooring. What must be the minimum dollar value of the certified wood used in order to comply with the requirements of LEED for New Construction MR Credit 7, Certified Wood?

    (A) $25,000
    (B) $50,000
    (C) $60,000
    (D) $75,000

LEED BD&C Practice Exam

31. A project team is seeking platinum certification for a brewery in Colorado and is evaluating sites to develop. Development at which locations would qualify the project for earning a point for SS Credit 1? (Choose three.)

    (A) within 40 feet (12 meters) of a small manufactured pond used for stormwater retention
    (B) within 200 feet (61 meters) of a small constructed wetland created to restore natural habitats
    (C) within 40 feet (12 meters) of a natural pond used in a geothermal heating/cooling source
    (D) within 50 feet (15 meters) of a small manufactured pond used for fire suppression
    (E) within 50 feet (15 meters) of a small manufactured pond used to restore natural habitats

32. A new 42,000 sq ft (3900 sq meter) daycare center will have a mechanical ventilation system. To meet the requirements of IEQ Credit 2, Increased Ventilation, the project will need to exceed ANSI/ASHRAE 62.1 standards by what percentage to earn a point in the LEED for New Construction rating system?

    (A) 30%
    (B) 40%
    (C) 50%
    (D) 60%

33. Naturally and mechanically ventilated building projects must comply with which standard to meet the requirements of IEQ Prerequisite 1, Minimum Indoor Air Quality Performance?

    (A) ANSI/ASHRAE 52.2
    (B) ANSI/ASHRAE 62.1
    (C) ANSI/ASHRAE/IESNA 90.1
    (D) ASHRAE 129

34. A project team demonstrating compliance with the requirements of the prescriptive option of IEQ Credit 8.1, Daylight and Views: Daylight must include which of the following factors in the appropriate calculations? (Choose three.)

    (A) FTE occupancy
    (B) floor area
    (C) window area
    (D) solar heat gain coefficient
    (E) visible light transmittance
    (F) U-factor

35. A sun path diagram can be used to determine which of the following?

    (A) visible light transmittance of glass
    (B) solar heat gain coefficient of glass
    (C) dimensions, locations, and angles of exterior shading devices
    (D) vision glazing U-factor

36. How many points can be earned for exemplary performance under the LEED for New Construction rating system?

    (A) 2 points
    (B) 3 points
    (C) 4 points
    (D) 5 points

37. Local zoning requires that 25% of a site excluding the building footprint must be reserved for open space. The total site area is just over 445,000 sq ft (41,300 sq meters). The development footprint is 245,000 sq ft (22,800 sq meters) and the building footprint is 45,000 sq ft (4200 sq meters). How much land must the project team dedicate to vegetated open space to earn a point for SS Credit 5.2, Site Development: Maximize Open Space?

    (A) 45,000 sq ft (4200 sq meters)
    (B) 125,000 sq ft (11,600 sq meters)
    (C) 200,000 sq ft (18,580 sq meters)
    (D) 245,000 sq ft (22,800 sq meters)

38. Which of the following is NOT true regarding wastewater aquatic system effluent?

    (A) It can be used in irrigation.
    (B) It is considered potable.
    (C) It can be used cooling towers.
    (D) It can be used in water closets.

39. Hiring which of the following commissioning agents would contribute toward the meeting of the requirements of EA Credit 3, Enhanced Commissioning? (Choose two.)

    (A) a qualified employee of the design firm involved in the project, if the individual is not directly responsible for primary design of the project.
    (B) a qualified employee of the owner
    (C) a qualified consultant to the owner
    (D) a qualified employee of the contractor, if the individual is not serving another role on the project.
    (E) a qualified construction manager working for a contractor involved with the project

40. A whole building energy simulation model can assist with documenting which of the following LEED credits and prerequisites? (Choose four.)

    (A) EA Credit 1, Optimize Energy Performance
    (B) EA Credit 2, On-site Renewable Energy
    (C) EA Credit 3, Enhanced Commissioning
    (D) EA Credit 4, Enhanced Refrigerant Management
    (E) EA Credit 5, Measurement and Verification
    (F) EA Credit 6, Green Power

41. Which of the following accurately describes the *U*-factor?

    (A) a measure of the ability of a surface material to reflect sunlight
    (B) the ratio of total light to incident light passing through a glazing surface divided by the amount of light striking the glazing surface
    (C) the fraction of solar radiation admitted through a window or skylight
    (D) a measure of heat transfer through glazing or some other product

42. Which of the following would help a project earn an Innovation in Design point for exemplary performance?

    (A) diverting 90% of the waste generated through the construction process from a landfill
    (B) developing a project within a density radius of 130,000 sq ft/acre (29,800 sq meter/hectare)
    (C) providing daylight in 90% of the regularly occupied spaces in a school
    (D) reducing interior water use by 40%

43. To meet the requirements of IEQ Credit 5, Indoor Chemical and Pollutant Source Control, a mechanically ventilated building must use new air filters that have a minimum efficiency reporting value (MERV) of which of the following?

    (A) 5
    (B) 8
    (C) 10
    (D) 13

44. A project team is retrofitting an existing 10,500 sq ft (975 sq meters) library and designing an addition. What is the maximum area that the addition can be to meet the requirements of MR Credit 1.1, Building Reuse: Maintain Existing Walls, Floors, and Roof?

    (A) 20,000 sq ft (1860 sq meters)
    (B) 21,000 sq ft (1950 sq meters)
    (C) 22,000 sq ft (2050 sq meters)
    (D) 63,000 sq ft (5850 sq meters)

45. Vegetation selection affects the landscape coefficient. What variables are necessary to calculate the landscape coefficient?

(A) species factor, project evapotranspiration rate, and density factor
(B) microclimate factor, species factor, and density factor
(C) project evapotranspiration rate, density factor, and species factor
(D) microclimate factor, climate zone, and species factor

46. IEQ Credit 4.3, Low-Emitting Materials: Flooring Systems uses the Carpet and Rug Institute's Green Label Plus program to measure VOCs. How are VOCs measured?

(A) as a percentage of VOCs by weight
(B) in grams per liter minus water and exempt compounds
(C) in micrograms per square meter per hour
(D) in micrograms per cubic meter of air

47. A project team is working on bicycle accommodations for a nonresidential LEED for New Construction project. Which of the following must the project team install to achieve SS Credit 4.2, Alternative Transportation: Bicycle Storage and Changing Rooms?

(A) covered bicycle storage for 15% of building occupants
(B) bike lanes extending to the end of the property in at least 2 directions without barriers
(C) secure storage within 200 yards (183 meters) of a building entrance for at least 5% of all peak period building users
(D) covered bicycle storage facilities for 20% or more of building occupants

48. A school building is being developed on a site with no local zoning requirements. The building footprint is 50,000 sq ft (4600 sq meters). What is the minimum area of vegetated open space needed to meet the requirements of SS Credit 5.2, Site Development: Maximize Open Space?

(A) 50,000 sq ft (4600 sq meters)
(B) 25,000 sq ft (2300 sq meters)
(C) 100,000 sq ft (9300 sq meters)
(D) 150,000 sq ft (14,000 sq meters)

49. A LEED for New Construction project team intends to provide incentives for building occupants to bicycle to work. The project site does not include shower or locker facilities, but there are showers available in a recreation center within 100 feet (30 meters) of the project building's bicycle storage units. Which of the following would the project team MOST likely use to help determine whether the shower facilities in the recreation center can be used to meet LEED credit requirements?

(A) credit interpretation ruling
(B) local zoning requirements
(C) Uniform Building Code
(D) case studies of LEED-certified buildings

50. What are the potential tradeoffs of complying with the requirements of IEQ Credit 2, Increased Ventilation for active ventilation systems? (Choose two.)

    (A) increased solar heat gain
    (B) increased initial costs due to increased HVAC capacity
    (C) increased content of volatile organic compounds in outdoor air
    (D) increased operating costs due to higher HVAC costs
    (E) decreased occupant comfort, well-being, and productivity

51. Ground source heat pumps will help to contribute to meeting the requirements of which of the following LEED credits and/or prerequisites? (Choose two.)

    (A) EA Prerequisite 1, Fundamental Commissioning of the Building Energy Systems
    (B) EA Prerequisite 2, Minimum Energy Performance
    (C) EA Credit 1, Optimize Energy Performance
    (D) EA Credit 6, Green Power
    (E) EA Credit 2, On-site Renewable Energy
    (F) EA Credit 4, Enhanced Refrigerant Management
    (G) EA Credit 5, Measurement and Verification

52. Based on requirements for compliance with IEQ Credit 1, Outdoor Air Delivery Monitoring, at what height above the finished floor should a carbon dioxide sensor be installed?

    (A) 30 inches (0.8 meters)
    (B) 60 inches (1.5 meters)
    (C) 90 inches (2.3 meters)
    (D) 120 inches (3.0 meters)

53. A project team is tasked with remodeling an existing 26,000 sq ft (2400 sq meter) building and adding 60,000 sq ft (5600 sq meter). The project team is trying to reuse as many materials as possible from the existing building. The lighting designer plans to reuse the building's existing doors by having them painted white and utilizing them as light shelves. The light shelves will be installed on the interior of the building to help distribute and control daylight. The use of the existing doors as light shelves will NOT apply to which of the following LEED credits? (Choose two.)

    (A) MR Credit 1, Building Reuse
    (B) MR Credit 2, Construction Waste Management
    (C) MR Credit 3, Materials Reuse
    (D) MR Credit 5, Regional Materials
    (E) EA Credit 1, Optimize Energy Performance
    (F) IEQ Credit 8.1, Daylight and Views: Daylight
    (G) IEQ Credit 8.2, Daylight and Views: Views

54. Which of the following would meet the requirements for IEQ Credit 6.1, Controllability of Systems: Lighting, in an office building? (Choose two.)

    (A) continuous dimming in all offices
    (B) automatic off switching in all offices
    (C) operable windows with interior light shelves in all offices
    (D) portable, plug in task lights at all work stations
    (E) occupancy sensors and light sensors at all workstations
    (F) bi-level switching in all conference rooms

55. Which of the following must be commissioned in order for a project to achieve LEED certification? (Choose four.)

    (A) building envelope
    (B) domestic hot water systems
    (C) HVAC & R systems
    (D) renewable energy systems
    (E) lighting and daylighting systems
    (F) power distribution systems
    (G) permanently wired electrical motors

56. A project team is considering pursuing EA Credit 5, Measure and Verification; however, the owners are concerned with the cost of meeing its requirements. Which of the following factors affect the cost of this credit? (Choose three.)

    (A) quantity and type of metering points
    (B) voltage needed for power distribution systems
    (C) commissioning services upon installation of the energy systems
    (D) duration and accuracy of metering activities
    (E) availability of existing data collection systems

57. Which of the following is NOT required to comply with the requirements of ANSI/ASHRAE/IESNA 90.1 in order to meet the requirements of EA Prerequisite 2?

    (A) building envelope
    (B) renewable energy systems
    (C) lighting and daylighting systems
    (D) permanently wired electrical motors

58. A LEED for New Construction project team is striving to meet the requirements of MR Credit 7, Certified Wood. Which of the following entities is NOT required to maintain a chain of custody (COC) certification?

    (A) transport company
    (B) supplier/manufacturer
    (C) vendor
    (D) end user

59. IEQ Credit 4.4, Low-Emitting Materials: Composite Wood and Agrifiber Products, deals most directly with which of the following products? (Choose three.)

(A) dimensional lumber
(B) plywood
(C) reclaimed hardwood flooring
(D) door cores
(E) particleboard

60. If a LEED for Core and Shell project is under construction and cannot use permanently installed grates, grilles, or slotted systems, which of the following is true regarding the project's eligibility for compliance with the requirements of IEQ Credit 5, Indoor Chemical and Pollutant Source Control?

(A) The project is eligible after the owners assign the task of monthly maintenance of roll-out walk off mats to building staff.
(B) The project is eligible after the owners contract weekly maintenance of roll-out walk off mats.
(C) The project is eligible after installing the entryway system on the interior side of the entrance to the building.
(D) The project is ineligible.

61. ASHRAE 55 provides an optional method for cooling and ventilating naturally ventilated spaces, which USGBC references for compliance with IEQ Credit 7.1, Thermal Comfort: Design. This method accounts for additional factors that are present in naturally ventilation systems. Which of the following is NOT considered independent of ventilation conditions?

(A) air speed
(B) clothing
(C) humidity
(D) mean monthly outdoor temperatures

62. A project team may use which of the following to determine the amount of green power needed to meet the requirements of EA Credit 6, Green Power? (Choose two.)

(A) the estimated annual building operating costs
(B) the building's estimated annual electricity consumption
(C) U.S. DOE's Commercial Building Energy Consumption Survey database
(D) U.S. EPA's Energy Star Portfolio Manager
(E) International Performance Measurement and Verification Protocol

63. Based on July conditions, the baseline case for a project sets the total water applied (TWA) at 100,000 gallons (380,000 liters). The design case TWA is 43,400 gallons (164,000 liters). The design case water reuse contribution is 3400 gallons (12,900 liters) of water from the rainwater catchment system. As a percentage, what is the building's water use reduction?

    (A) 40%
    (B) 50%
    (C) 60%
    (D) 70%

64. Which of the following design changes would contribute toward meeting the requirements of IEQ Credit 6.2, Controllability of Systems: Thermal Comfort?

    (A) ensuring that the area of window opening be equal to 3% of net occupiable area
    (B) installing operable windows according to ASHRAE 55
    (C) installing thermostats in conference rooms
    (D) installing occupancy sensors to turn off heating and cooling

65. A project is being constructed on an existing parking lot. The pre-development site runoff rate is 0.80 cu ft per second and the pre-development site runoff quantity is 10,000 cu ft. Which of the following reflects the minimum *post*-development site runoff rate and quantity needed to earn a point under SS Credit 6.1, Storm water Management: Quantity Control?

    (A) rate is 0.80 cu ft per second (0.022 cu meters per second); quantity is 10,000 cu ft (280 cu meters)
    (B) rate is 0.50 cu ft per second (0.014 cu meters per second); quantity is 6,250 cu ft (180 cu meters)
    (C) rate is 0.40 cu ft per second (0.11 cu meters per second); quantity is 5,000 cu ft (140 cu meters)
    (D) rate is 0.18 cu ft per second (0.01 cu meters per second); quantity is 2,250 cu ft (60 cu meters)

66. IEQ Prerequisite 2, Environmental Tobacco Smoke (ETS) Control and IEQ Credit 5, Indoor Chemical Pollutant Source Control require a pressure difference in rooms relative to the surrounding areas. When the doors are closed, what is this required pressure difference?

    (A) an average of 3 Pa and a minimum of 1 Pa
    (B) an average of 5 Pa and a minimum of 1 Pa
    (C) an average of 13 Pa and a minimum of 3 Pa
    (D) an average of 25 Pa and a minimum of 15 Pa

67. An urban project with minimal open space is using rainwater cisterns to manage and treat stormwater. To comply with the requirements of SS Credit 6.2, Stormwater Design: Quality Control, the civil engineer must verify that the cisterns can do which of the following?

(A) infiltrate at least 80% of the annual rainfall volume
(B) infiltrate at least 90% of the annual rainfall volume
(C) accommodate at least 80% of the annual rainfall volume
(D) accommodate at least 90% of the annual rainfall volume

68. Which of the following is NOT a potential tradeoff of compliance with the requirements of IEQ Credit 3.2, Construction Indoor Air Quality Management Plan: Before Occupancy?

(A) increased energy use
(B) increased contaminants from outside air
(C) increased cost for labor and time
(D) increased cost for air filtration media

69. A project architect is planning to use an existing bridge as a source of structural steel. The steel will need to be sandblasted and painted to remove and/or seal any lead paint that was used in the original installation. The bridge is located about 180 miles (290 kilometers) from the project site. Which of the following LEED credits will the reuse of the steel apply towards? (Choose two.)

(A) MR Credit 1, Building Reuse
(B) MR Credit 2, Construction Waste Management
(C) MR Credit 3, Material Reuse
(D) MR Credit 4, Recycled Content
(E) MR Credit 5, Regional Materials
(F) MR Credit 6, Rapidly Renewable Materials

70. A LEED for New Construction project team working on a commercial office building would like to earn MR Credit 6, Rapidly Renewable Materials. Using the 45% default materials value, it has determined the total value of materials to be $1,000,000 based on the hard costs in CSI Masterformat 2004 Division 3-10, 31, and 32 only. The team has specified that the project will use $10,000 worth of Forest Stewardship Council-certified lumber, $15,000 worth of reclaimed wood floors, and $15,000 worth of bamboo flooring. Which of the following is the minimum value of additional rapidly renewable materials necessary in order to meet the requirements of both MR Credit 6, as well as one point for exemplary performance?

(A) $0
(B) $20,000
(C) $25,000
(D) $35,000

71. A project team would like to pursue Option 1 for EA Credit 1, Optimize Energy Performance, which is worth up to 19 points and requires an energy simulation model using the performance rating method. In the calculations for compliance with this credit's requirements, which of the following values is compared to the baseline case?

    (A) annual electricity consumption
    (B) annual energy consumption
    (C) annual energy costs
    (D) annual cost of renewable energy credits purchased to offset energy consumption

72. When must the computer simulation calculation be performed to demonstrate compliance with the requirements of the simulation option for IEQ Credit 8.1, Daylight and Views: Daylight?

    (A) April 21 at 9 am and 3 pm
    (B) June 21 at 9 am and 3 pm
    (C) September 21 at 9 am and 3 pm
    (D) December 21 at 9 am and 3 pm

73. An architectural team is touring an existing building to determine which components can be reused for a new project. Which of the following materials are excluded from the calculations in MR Credit 1, Building Reuse? (Choose three.)

    (A) window assemblies
    (B) roof decks
    (C) exterior walls
    (D) interior walls
    (E) structural supports
    (F) nonstructural roofing material
    (G) foundations

74. EA Credit 1, Optimize Energy Performance, provides four options for compliance, the second of which is based on ASHRAE's Advanced Energy Design Guides. Which of the following projects would qualify to follow Option 2 for credit compliance? (Choose two.)

    (A) 15,000 sq ft (1400 sq meter) retail building
    (B) 25,000 sq ft (2300 sq meter) office building
    (C) 45,000 sq ft (4200 sq meter) warehouse
    (D) 100,000 sq ft (9300 sq meter) self storage building
    (E) 225,000 sq ft (21,000 sq meter) K-12 school

75. Which of the following are basic systems used to ventilate buildings? (Choose three.)

    (A) electrical ventilation systems
    (B) hydronic ventilation systems
    (C) mechanical ventilation systems
    (D) natural ventilation systems
    (E) mixed mode ventilation systems
    (F) solar ventilation systems

76. Installing which of the following filters will contribute toward compliance with the requirements of IEQ Credit 3, Construction Indoor Air Quality Management Plan?

    (A) MERV 8 filters
    (B) MERV 13 filters
    (C) HEPA 8 filters
    (D) HEPA 13 filters

77. A project team designing a 104,000 sq ft (970 sq meter) warehouse in Tucson, Arizona with a mechanical HVAC system is considering complying with the requirements of IEQ Credit 2, Increased Ventilation. The warehouse is designed with a mechanical HVAC system. Which of the following can be used to estimate the additional operating costs that will be incurred by increasing ventilation?

    (A) whole building energy simulation model
    (B) daylight simulation model
    (C) fluid thermodynamics model
    (D) none of the above

78. Which of the following accurately defines solar heat gain?

    (A) measure of the ability of a surface material to reflect sunlight
    (B) ratio of total light to incident light passing through a glazing surface divided by the amount of light striking the glazing surface
    (C) fraction of solar radiation admitted through a window or skylight
    (D) measure of heat transfer through glazing or some other product

Practice Exam Part Two

79. A project site is classified as being in a "LZ2: Low" lighting zone under SS Credit 8, Light Pollution Reduction. Currently, 3% of the total initial designed fixture lumens, based on the sum of all fixtures on site, is emitted an angle above 90 degrees or higher from nadir. There are three different types of lighting on site. One includes luminaires that meet the full cutoff classification in the parking lot areas and are mounted on the building. There are also luminaires that meet the cutoff classification used in pedestrian areas. What action, if any action is needed, could the project team take to reduce the amount of uplight emitted by the exterior lighting systems? (Choose two.)

   (A) No action is required in order to comply with LZ2.
   (B) Provide full cutoff luminaires in the pedestrian areas.
   (C) Reduce the wattage of the building-mounted lamps.
   (D) Reduce the wattage of the lamps in the pedestrian areas.
   (E) Provide semi-cutoff luminaires for the parking lot lighting.

80. A project team would like to meet the requirements of IEQ Credit 3.1, Construction IAQ Management Plan: During Construction, which references the Sheet Metal and Air Conditioning Contractors' of North America (SMACNA) IAQ Guidelines for Occupied Buildings Under Construction. These guidelines recommend control measures for which of the following? (Choose three.)

   (A) HVAC protection
   (B) entryway systems
   (C) pathway interruption
   (D) environmental tobacco smoke control
   (E) scheduling

81. One year after a project's completion, the mechanical engineer and facilities manager for the project are completing a plan for corrective action in accordance with the requirements of IEQ Credit 7.2 Thermal Comfort: Verification. Corrective actions for which of the following should NOT be included in the plan?

   (A) diffuser airflow
   (B) solar control
   (C) occupancy sensors
   (D) maintenance scheduling

82. A non-residential LEED for New Construction project's parking capacity does not exceed local zoning requirements. If the parking lot has 100 total spaces, how many preferred parking spaces for carpools or vanpools must the project team provide to fully comply with the remaining requirement for SS Credit 4.4: Alternative Transportation: Parking Capacity's Option 1?

   (A) 3 spaces
   (B) 5 spaces
   (C) 10 spaces
   (D) 15 spaces

83. A project team is evaluating a previously undeveloped site upon which to construct a 40,000 sq ft (3700 sq meter) office building. Development on which of the following sites would prohibit a project from earning SS Credit 1, Site Selection?

    (A) a site 5 feet (1.5 meters) above a 100 year flood plane
    (B) a site 55 feet (17 meters) from a wetland
    (C) a site 55 feet (17 meters) from a body of water
    (D) brownfield site

84. A project team is preparing a new school's master plan for review by the school board. Sharing which of the following facilities with the community will contribute toward meeting the requirements of SS Credit 10, Joint Use of Facilities? (Choose two.)

    (A) healthcare center
    (B) gymnasium
    (C) media center
    (D) administrative office
    (E) parking lot

85. The project team for a proposed school plans to conduct a Phase I Environmental Site Assessment to determine if a site is contaminated. Which of the following would guarantee LEED certification ineligibility?

    (A) The environmental assessment is performed by a member of the project team.
    (B) Historical records related to the site are reviewed.
    (C) Neighbors of the proposed site state the land was previously used as a landfill.
    (D) Critical levels of contamination are found.

86. Installing which of the following can contribute toward compliance with the requirements of WE Credit 4, Process Water Use Reduction?

    (A) a dishwasher that uses no more than 1 gallon (3.8 liters) of water per rack
    (B) refrigeration equipment that uses once-through cooling
    (C) process equipment that achieves 15% reduction in water use over industry standards
    (D) a garbage disposal

87. Which of the following is NOT needed to perform calculations within Water Efficiency Credit 3, Water Use Reduction?

    (A) usage group definitions
    (B) number of occupants
    (C) type of occupants
    (D) installed annual water use rates

88. For permanently installed wood products, project teams must collect all of the following EXCEPT which from the vendor?

    (A) proof of FSC certification for each wood product purchased
    (B) cost of each wood product purchased
    (C) line-item list of wood products purchased
    (D) COC certificate number on invoices with FSC products

89. Which of the following is considered to be a *regularly occupied space* in a school building?

    (A) gymnasium
    (B) auditorium
    (C) locker room
    (D) main lobby

90. Teachers and school staff are working with the project team on the design of a new charter school. Productivity, well-being, and effective learning are the primary objectives of the new school. What strategy for background noise and reverberation reduction is appropriate?

    (A) installing low-NRC acoustical tiles
    (B) implementing an open classroom plan with sliding walls that separate instructional areas
    (C) mounting high-NRC acoustical tiles with the minimum possible air space behind tiles
    (D) installing storm windows

91. LEED for Schools project teams must meet the requirements of which of the following credits to comply with the requirements of LEED for Schools' IEQ Credit 10, Mold Prevention?

    (A) IEQ Credit 2, Increased Ventilation
    (B) IEQ Credit 3.1, Construction Indoor Air Quality Management Plan: During Construction
    (C) IEQ Credit 3.2, Construction Indoor Air Quality Management Plan: Before Occupancy
    (D) IEQ Credit 5, Indoor Chemical and Pollutant Source Control

92. Which of the following is the minimum indoor noise level that could contribute toward a LEED for Schools project team's earning an exemplary performance point for IEQ Credit 9, Enhanced Acoustical Performance?

    (A) 35 dBA
    (B) 40 dBA
    (C) 45 dBA
    (D) 50 dBA

93. ANSI/ASHRAE 62.1 sets ventilation rate requirements to mitigate people-related source contaminants and area-related source contaminants. Mitigating area-related source contaminants will improve which of the following?

    (A) actual occupancy density
    (B) background off gassing
    (C) actual occupancy activity
    (D) system-level outdoor air

94. In the context of integrated design, which of the following statements are true about design charrettes? (Choose two.)

    (A) Design charrettes reveal potential community alliances and partnerships.
    (B) Design charrettes reveal synergies and tradeoffs among green building strategies.
    (C) Design charrettes are led by the LEED AP.
    (D) Only the project owner and project team can be involved in the design charrette.
    (E) The project team is required to complete a minimum of one design charrette during the LEED certification process.

95. To comply with the requirements of IEQ Credit 7.2, Thermal Comfort: Verification, the owners should use which of the following referenced standards of environmental variables in potential problem areas?

    (A) ANSI/ASHRAE 52.2
    (B) ANSI/ASHRAE 62.1
    (C) ANSI/ASHRAE/IESNA 90.1
    (D) ASHRAE 55

96. A project team is designing the site plan for a project being built on a greenfield. Site disturbances should be limited to what distance to comply with the requirements of SS Credit 5.1, Site Development: Protect or Restore Habitat?

    (A) 15 feet (4.5 meters) beyond primary roadway curbs and main utility branch trenches
    (B) 15 feet (4.5 meters) beyond surface walkways, patios, surface parking, and utilities less than 12 inches (0.3 meters) in diameter
    (C) 30 feet (9.1 meters) beyond the building perimeter
    (D) 40 feet (12 meters) beyond constructed areas with permeable surfaces

97. The project boundary must be used consistently in LEED BD&C rating systems' calculations. In which of the following credits is the project boundary referenced? (Choose three.)

    (A) SS Prerequisite 1, Construction Activity Pollution Prevention
    (B) SS Credit 1, Site Selection
    (C) SS Credit 2, Development Density & Community Connectivity
    (D) SS Credit 5, Site Development
    (E) SS Credit 6, Stormwater Management
    (F) SS Credit 8, Light Pollution Reduction

98. Which of the following are considered biofuels for EA Credit 2, On-site Renewable Energy? (Choose three.)

    (A) untreated wood waste, including mill residues
    (B) natural gas
    (C) agricultural crops or waste
    (D) landfill gas
    (E) petroleum gas
    (F) output from a vapor compression system
    (G) animal waste or other organic waste

99. A school is being constructed in an urban area and the project team is attempting to qualify for a development density credit. Which of the following can be excluded as part of the project site from the development density calculations? (Choose three.)

    (A) turf-grass fields for physical education
    (B) public roads
    (C) concession stands used only during school sporting events
    (D) parking lots
    (E) total area of the building

100. A grade school is being constructed in an urban area and the project team is attempting to meet the requirements of SS Credit 4.1, Alternative Transportation: Public Transportation. In order to do so, what percentage of the students in grades eight and below must live within a 0.75 mile (1200 meter) radius of the school?

    (A) 60%
    (B) 70%
    (C) 80%
    (D) 90%

Practice Exam Part One Solutions

Solutions begin on the page that follows.

LEED BD&C Practice Exam

1. B
2. B
3. C
4. B
5. D
6. A
7. A
8. C
9. A
10. C
11. C
12. C
13. B, E
14. C
15. B
16. C
17. A
18. B
19. B
20. C, D
21. A, C, E
22. B
23. B
24. D
25. C
26. C
27. A
28. B, C, E
29. B
30. A, D, E
31. C
32. D
33. C
34. A
35. B
36. A
37. C
38. C
39. A
40. B
41. D
42. B
43. B
44. C
45. A, C, E
46. B, C, D, E, F
47. B, C
48. C
49. B
50. A
51. A
52. A
53. D, E, F
54. B
55. B
56. B, D
57. C
58. A
59. A, B, E
60. C
61. C
62. B
63. D
64. D
65. B
66. A, B, E
67. A
68. B
69. B
70. D
71. B
72. C
73. C
74. C
75. B
76. B, D, F
77. C, D, F
78. C
79. A
80. B
81. B
82. A
83. B, E
84. C, E
85. B, F
86. A, C
87. C
88. C, D, F
89. B, C
90. A
91. A, B
92. A
93. C
94. C
95. C, D
96. A, E, F
97. A
98. D
99. A
100. B, E

Practice Exam Part One Solutions

1. *The answer is:* (B) demonstrating potable water use reductions by water-consuming fixtures

A mechanical engineer is responsible for demonstrating the reduction of potable water use by a building's interior water-consuming fixtures. A civil engineer is responsible for declaring compliance with either NPDES or local control standards, demonstrating that stormwater management strategies reduce stormwater runoff, and listing structural controls with pollutant removal methods and percent annual rainfall treated.

2. *The answer is:* (B) civil engineer

According to the U.S. EPA, a brownfield site is real property whose expansion, redevelopment, and/or reuse may be complicated by the presence or potential presence of a hazardous substance, pollutant, or contaminant. A civil engineer does not have the authority to classify a LEED site as a brownfield. Project teams can earn a point for developing on a brownfield only if the site has been classified as such by a local, state, or federal agency, or as contaminated by means of the ASTM E1903-97 Phase II Environmental Site Assessment.

3. *The answer is:* (C) American National Standards Institute

The purpose of American National Standards Institute (ANSI) is to help U.S. businesses increase global competitiveness and to enhance the quality of life by promoting, facilitating, and safeguarding the integrity of voluntary consensus standards and conformity assessment systems. Acoustical standards are set for LEED for Schools projects within ANSI S12.60, *Acoustical Performance Criteria, Design Requirements and Guidelines for Schools.*

4. *The answer is:* (B) simultaneously

While projects may earn multiple LEED certifications, project teams may not pursue multiple LEED certifications simultaneously. If more than one LEED rating system applies to a particular project, the project team must decide which one to pursue first. For example, a project that becomes certified under one of the LEED Building Design and Construction rating systems can also become certified under the LEED for Existing Buildings: Operation and Maintenance rating system at a later date.

5. *The answer is:* (D) ventilation

ANSI/ASHRAE 62.1 specifies minimum ventilation rates and acceptable indoor air quality levels, thereby minimizing the potential for adverse health effects. This standard is referred to in the Indoor Environmental Quality credit category.

6. *The answer is:* (A) building floor area

To determine the baseline water calculation for potable use, project teams must document the types of water-using fixtures and the frequency-of-use data, including the number of full-time equivalent occupants, the duration of the fixture's use, and the water volume per fixture use. The building floor area (i.e., building square footage) is not needed to determine the baseline usage.

7. *The answer is:* (A) radiation emitted by a surface to the radiation emitted by a black body

The LEED reference guides define emissivity as the ratio of radiation emitted by a surface to the radiation emitted by a blackbody at the same temperature. It is also the ability of a material to shed infrared radiation or heat. Emissivity characteristics and the solar reflectance index (SRI) are used to determine the compliance of roofing materials with the requirements of the heat island effect credits in the LEED rating systems.

8. *The answer is:* (C) renovation of a tenant space from a hair salon into a coffee shop

The LEED rating systems have evolved to address the various market sectors of the building industry and include new construction, tenant finish activities, existing buildings, core and shell development, residential construction, multi-building campuses, retail, and several others in development such as laboratories, health care, and neighborhood development. The LEED for Commercial Interiors rating system is the most applicable to an interior renovation, such as a tenant space changing from a hair salon to a coffee shop, because it gives a business leasing a tenant space the opportunity to become LEED-certified.

The construction of a 9000 sq ft (840 sq meter) commercial real estate office building on a previously impacted site would be eligible for LEED certification under the New Construction rating system. The retrofit of the mechanical and electrical systems of an existing commercial tenant space mainly deals with ongoing operation and maintenance of a building and therefore would be certified under the LEED for Existing Buildings: Operations and Maintenance rating system. The construction of an office complex for future commercial tenants would be best certified under the LEED for Core and Shell rating system, where it allows developers to provide the building structure and base systems for future tenants.

9. *The answer is:* (A) using recycled-content building materials

The LEED rating systems allow project teams to include flyash as a pre-consumer recycled-content material. Flyash is a waste material from coal-fired power plants and can partially replace cement in certain concrete applications. Varying percentages of flyash can be used in the concrete mix, depending on the building application. The structural engineer will specify the concrete mix based on the structural needs of the building application. Though flyash has other applications, if it is not used in concrete, it is generally sent to a landfill.

10. *The answers is:* (C) date of submission for review

The certification fee varies based on the rating system under which the project is becoming LEED certified and the size of the project. USGBC members pay lower fees than non-members. The certification fee is paid when the project team submits documentation for review via LEED Online; however, the submission date does not affect the overall fee.

11. *The answer is:* (C) CFCs

The U.S. government banned CFC production in 1995. In the Energy and Atmosphere credit category, USGBC requires that HVAC & R base-building and fire-suppression systems contain no chlorofluorocarbons (CFCs). Though the federal government has banned the use of CFCs in most applications, the LEED rating systems and their applicable reference guides also have set phase-out parameters for buildings that contain old CFC-based HVAC systems.

Practice Exam Part One Solutions

12. The answer is: (C) a situation in which pursuing multiple strategies requires a compromise

Tradeoffs, as related to the LEED rating systems, occur when a particular green building strategy has a negative effect on another green building strategy. For example, increasing the ventilation rates in a building can increase building occupant productivity and result in LEED credit-earning; however, this increased level of ventilation may also increase the amount of energy consumed by HVAC systems and make LEED credit-earning in the Energy and Atmosphere credit category more difficult.

13. The answers are: (B) outdoor air delivery monitoring systems
(E) refrigerant management

The mechanical engineer plays a major role in the documentation of many of the LEED Indoor Environmental Quality credits. These credits relate to the HVAC systems and include parameters for ventilation rates and effectiveness, outdoor air introduced in the building, and thermal comfort. Stormwater management systems are designed by the civil engineer. The architect and general contractor are generally responsible for specifying and installing low emitting materials such as paints, adhesives, sealants, carpets, and composite woods.

14. The answer is: (B) dual-flush toilets

Installing dual-flush toilets will reduce water consumption in a building. Automated toilet sensors are installed for sanitation and will not reduce water consumption. In fact, automated toilet sensors often increase water consumption due to false and/or unnecessary flushes. Pressure assisted and blow-out toilets are offered in low flow versions, but not all makes and models will reduce water consumption.

15. The answer is: (B) complete the online registration

Completing GBCI's online registration is the first step toward LEED certification. The precertification application is part of the LEED for Core and Shell process only, and it is submitted after project registration. There is no requirement to include a LEED AP on the project team, although teams can earn a point when one is included as a primary member of the project team.

16. The answer is: (C) when GBCI performs an audit

GBCI does not require material safety data sheets (MSDS) or cut sheets as part of the typical submittal process. However, GBCI can audit several credits as part of the review and certification process, and at that time it may be necessary to submit MSDS or cut sheets and additional information. Materials with low volatile organic compounds (VOCs) are referenced in the Indoor Environmental Quality category of the LEED rating systems.

17. The answer is: (A) gas-electric hybrid car with a green score of 30 from the American Council for an Energy Efficient Economy vehicle rating guide

USGBC defines low-emitting and fuel-efficient vehicles as vehicles with a minimum green score of 40 on the American Council for an Energy Efficient Economy vehicle rating guide or as vehicles classified as zero emission vehicles (ZEV) by the California Air Resources Board.

18. *The answer is:* (B) plumbing fixtures

While the 1992 Energy Policy Act addresses standards for energy use, it also provides guidelines for plumbing fixtures and water conservation. LEED uses this standard to establish baseline water consumption and flow requirements for plumbing fixtures.

19. *The answer is:* (B) sustainable design guide and user's manual for the LEED for New Construction, Core and Shell, and Schools rating systems

The LEED BD&C reference guide is the definitive resource of the LEED for New Construction (NC), Core and Shell (CS), and Schools rating systems. It includes guidance on LEED design and construction; provides sample calculations, diagrams, and designs; and offers perspectives on the environmental issues related to the green building strategies and technologies addressed by each credit and prerequisite within the LEED rating systems.

20. *The answers are:* (C) market exposure through the USGBC website
(E) third-party validation of a building's performance

LEED certified projects enjoy the benefit of market exposure through the USGBC website. Third-party validation of their building's performance, rather than self-validation, is another benefit of LEED certification. USGBC offers discounts to its member companies, but it doesn't offer discounts on products or services to owners or members of LEED certified projects.

21. *The answers are:* (A) case studies of LEED projects
(C) searchable list of LEED APs

USGBC uses its website and LEED Online as a primary means of communicating with the green building industry nationwide. Therefore, its website contains case studies of LEED projects and a searchable list of LEED APs.

Anyone can sign up for an account with USGBC. This account can help manage projects and establish communication with USGBC. There are additional resources and tools at the GBCI website.

22. *The answers are:* (B) humidity
(E) temperature

ASHRAE 55, *Thermal Environmental Conditions for Human Occupancy*, focuses on environmental factors such as temperature, humidity, and air speed, and personal factors such as activity and clothing. Meeting the requirements of ASHRAE 55 will, in most cases, result in thermally comfortable conditions acceptable to 80% or more of the occupants within a given space.

23. *The answer is:* (B) post-consumer recycled content

USGBC refers to two primary types of recycled content: *post-consumer* and *pre-consumer*. USGBC defines *post-consumer recycled content* as material generated by households or by commercial, industrial, and institutional facilities in their role as end-users of a product that can no longer be used for its intended purpose. An example of a material made with post-consumer recycled content is carpet made from beverage bottles that were used in a home and then recycled. *Pre-consumer recycled content* is industrial waste that becomes feedstock for another industrial process. USGBC does not consider industrial waste that becomes feedstock in the same process to be recycled content. An example of a material with pre-consumer (sometimes called post-industrial) recycled content is carpet that is made from plastic waste from a beverage container manufacturer.

USGBC considers post-consumer recycled content to be more sustainable than pre-consumer recycled content (and therefore provides a higher weighted average for post-consumer recycled content) because post-consumer recycled materials have reached the end user and therefore have received more use from the embodied energy that goes into the materials. It is also more difficult for a material to find its way back into an industrial process after the consumer has used it.

24. The answer is: (D) not using refrigerants

Project teams can earn points in the Energy and Atmosphere credit category for not using refrigerants. Meeting the minimum requirements of ANSI/ASHRAE 62.1's ventilation rate procedure and approved addenda, and ANSI/ASHRAE/IESNA 90.1 are prerequisites under the LEED rating systems. Prerequisites do not themselves earn a project points in the rating system, but a project can only achieve LEED certification if all prerequisites are met. Points are earned when projects meet or exceed the requirements specified in LEED credits.

25. The answer is: (C) ongoing measurement and verification

For each proposed Innovation in Design credit, the project must identify the credit's intent, the team's proposed requirements for compliance, submittals to demonstrate compliance, and strategies that might be used to meet the requirements. There is currently no requirement to document ongoing measurement and verification. This concept is addressed in the EA category of the LEED rating systems.

26. The answer is: (C) 1.0

The number of full-time equivalent (FTE) building occupants is equal to the number of worker hours divided by eight hours. (Eight hours represents a full-time workday.) According to USGBC, a full-time worker has an FTE value of 1 person, and a half-time worker has an FTE value of 0.5 person.

27. The answer is: (A) energy performance of buildings

ANSI/ASHRAE/IESNA 90.1 sets minimum standards for energy-efficient design of buildings. Specifically, it refers to regulated building loads such as building envelope, HVAC, service water heating, power, lighting, and other equipment, including all permanently wired electrical motors.

LEED BD&C Practice Exam

28. *The answers are:* (B) mulching

(C) permanent seeding

Permanent seeding, temporary seeding, and mulching are all examples of stabilization control measures, or planting techniques that stabilize the soil and minimize erosion and sedimentation. Earth dikes, silt fences, sediment traps, and sediment basins are examples of structural control measures, or the use of structures to control erosion and sedimentation.

29. *The answer is:* (B) air filtration

The minimum efficiency reporting value (MERV) is based on a filter's ability to remove particles and the filter's resistance to airflow. This rating comes from ANSI/ASHRAE 52.2, which provides a method for testing the performance of air filters. MERV filter ratings are referred to in the LEED Indoor Environmental Quality credit category.

30. *The answers are:* (A) designated smoking areas, if smoking is not prohibited

(D) project team's intention to meet only the minimum local zoning requirements for parking

It is the responsibility of the project owner to declare the building's smoking policy if smoking is not prohibited, as well as the project team's intention regarding local zoning requirements for parking.

The mechanical engineer is responsible for decisions regarding installation, operational design, and controls or zones for the carbon dioxide monitoring system, as well as establishing the baseline for the water use calculations. The architect is responsible for deciding the location of the recycling area.

31. *The answer is:* (C) monitor outdoor air delivery

Outdoor air delivery monitoring is an Indoor Environmental Quality credit, and it is not a prerequisite for any LEED rating system.

32. *The answer is:* (D) project administrator

The project administrator is responsible for assigning roles to project team members. Once team members have been assigned roles, they can submit and have online access to their corresponding letter templates. More information on managing team roles can be found in the Team Admin area of the help section of LEED Online.

33. *The answer is:* (C) SCAQMD

The South Coast Air Quality Management District (SCAQMD) encourages minimizing air pollution and its impacts on public health. In the context of LEED, SCAQMD guidelines encourage better air quality in home and work environments. SCAQMD Rule 1168, which sets VOC limits for sealants used as fillers and for architectural coatings, is referenced in the Indoor Environmental Quality credit category.

34. *The answer is:* **(A)** Sustainable Sites

The LEED rating systems address potential negative development impacts on habitats through the Sustainable Sites credits that relate to site development and site disturbance. These credits require that projects on greenfield sites maintain development boundaries to protect open space, and a strategy to avoid site disturbance is to set up construction staging areas.

35. *The answer is:* **(B)** ratio of reflected solar energy to incident solar energy

Solar reflectance, or albedo, is defined as the ability of a surface material to reflect sunlight and is the ratio of reflected solar energy to incoming solar energy, expressed as a number between 0 and 1. Higher values of solar reflectance result in better control of heat gain. Solar reflectance includes the visible light, infrared heat, and ultraviolet wavelengths.

Emissivity is commonly mistaken for solar reflectance. They have similar characteristics, but emissivity pertains to the ability of a surface to emit or shed infrared heat.

36. *The answer is:* **(A)** LEED scorecard

Some LEED scorecards can be found on the USGBC website under the Certified Project List, but scorecards are not included with LEED case studies. LEED case studies provide overviews of LEED-certified projects, highlight green building strategies, and may provide general cost data, areas, project team information, and pictures. Additionally, they give insight on projects that have successfully implemented green building strategies based on factors such as region, construction type, building type, and project size.

37. *The answer is:* **(C)** access to the credit interpretation rulings database

Upon registration, LEED project teams gain access to software tools, LEED Online, and the credit interpretation rulings database. Access to this information early in a project (such as during the programming and schematic phases) allows project teams to better plan their LEED strategies, thus maximizing the potential for LEED certification.

Technical Advisory Groups (TAGs) only respond to credit interpretation requests, not to general email. The credit interpretation requests can be submitted to USGBC for a fee. Project teams do not get any discount on certification fees for registering a project early.

38. *The answer is:* **(C)** carpeting made from recycled plastic bottles

USGBC refers to two primary types of recycled content: *post-consumer* and *pre-consumer*. Pre-consumer recycled content is industrial waste that becomes feedstock for another industrial process, which would include all answer options except carpeting made from recycled plastic bottles (a post-consumer recycled material). Industrial waste that becomes feedstock in the same process is not considered recycled content according to USGBC.

USGBC considers post-consumer recycled content to be more sustainable than pre-consumer recycled content (and therefore provide a higher weighted average for post-consumer recycled content). Post-consumer recycled materials have reached the end user and therefore have received more use from the embodied energy that goes into the materials.

39. *The answer is:* (A) hard costs only, hard costs only

The total materials cost may be calculated as a percentage of the total construction cost based on the hard costs only in CSI Masterformat 2004 Divisions 3-10, 31, and 32, or as the sum of actual material costs based on the hard costs only in CSI Masterformat 2004 Divisions 3-10, 31, and 32. This figure represents the cost of building materials only. It does not include influences such as overhead, profit, rental fees, permit fees, soft costs, or labor. It also excludes electrical and mechanical systems. (These systems can be included as an Innovation in Design.) Any furnishings that are included must be included consistently throughout the material calculations.

40. *The answer is:* (C) bathroom sinks

Graywater systems are becoming more common as water demands are becoming a higher priority on the international level. Graywater can be captured from rainwater and sinks and then filtered and used for site irrigation, toilet flushing, or other non-potable water applications. Graywater sources include water from lavatory sinks, showers, bathtubs, and washing machines. All other possible options, along with any source of wastewater containing organic, toxic, or hazardous materials, are considered blackwater, which is also known as raw sewage and requires significant treatment before it can be reused.

41. *The answer is:* (D) mechanical engineer and general contractor

The mechanical engineer typically provides an indoor air quality management plan that incorporates and/or references the Sheet Metal and Air Conditioning Contractors' National Association (SMACNA) guidelines. The general contractor and subcontractors must comply with these guidelines during construction. USGBC references the SMACNA guidelines in the Indoor Environmental Quality Category. Note there are other requirements for meeting this credit in addition to SMACNA guidelines.

42. *The answer is:* (B) Green Power

To comply with the Green Power credit, USGBC requires the owner or responsible party to purchase, for a two-year period, a minimum of 35% of the building's power from a Green-e renewable source as defined by the Center for Resource Solutions (CRS) Green-e product certification requirements. The Renewable Energy credit relates to power created on a project's site, not power purchased from a Green-e renewable source. The Optimize Energy Performance credit deals with energy efficiency strategies and harvesting free energy. The Rapidly Renewable Resources credit deals with those materials that can be harvested in a 10-year life cycle or less.

43. *The answer is:* (B) 30 uses at 1.6 gallons (6.1 liters)

According to USGBC, a full-time equivalent (FTE) female occupant uses a conventional water closet three times each day. Therefore, 10 females make up 30 uses in one day. A conventional water closet uses 1.6 gallons (6.1 liters) per flush. The LEED reference guides provide tables with occupant water use default values and a table on Energy Policy Act of 1992 water use requirements. These tables provide the data for determining the water use reduction from baseline standards.

44. *The answer is:* (C) The CIR and ruling must be submitted with the LEED application.

The credit interpretation request and ruling process was established for project applicants seeking technical and administrative guidance on how LEED credits apply to their projects and vice versa. The CIR and ruling must be submitted with the LEED application. A ruling on a credit interpretation request does not guarantee a credit award, as the LEED certification application must still demonstrate and document achievement of credit requirements in order for credit to be awarded. Credit interpretation requests must not be formatted as a letter. Only the inquiry and essential background and/or supporting information that provide relevant project details should be included.

45. *The answers are:* (B) second level appeals

(D) rulings on credit interpretation requests

Project teams can appeal a first level appeal decision, but not a second (final) level appeal decision. Project teams can also not officially appeal a ruling on a credit interpretation request, although there may be a request for additional information. Decisions regarding minimum program requirements, credits, and prerequisites can be appealed through the appeals process, which is described on the GBCI website.

46. *The answers are:* (B) Innovation in Design

(C) Innovation in Operations

There are seven credit categories in the LEED BD&C rating systems: Sustainable Sites, Water Efficiency, Energy and Atmosphere, Materials and Resources, Indoor Environmental Quality, Innovation in Design (Innovation in Operations for LEED for Existing Buildings: Operations and Maintenance), and Regional Priority.

Under the Innovation in Design (or Innovation in Operations) credit category, project teams can earn points in two ways. The first is for exemplary performance, where teams earn points for exceeding the credit requirements by a percentage established by the rating system. The second is for innovation, where teams can earn points for implementing innovative strategies not specifically addressed by the particular project's applicable rating system.

Regional Priority credits provide incentives for the achievement of credits addressing regional environmental issues. Measurement and Verification is a credit in the Energy and Atmosphere credit category.

47. *The answers are:* (C) irrigation systems

(E) plumbing systems

USGBC defines *process water* as water used for industrial processes and building systems, such as cooling towers, boilers, and chillers. The term can also be used to refer to water used in operational processes, such as dishwashing, clothes washing, and ice making.

Process water is not included in the water reduction calculations for credit earning in most LEED rating systems, but process water loads can be included in exemplary performance calculations for water use reduction. LEED for Schools, however, does have a credit that deals specifically with process water.

48. *The answer is:* (C) furniture

In the Materials and Resources credit category, LEED project teams must calculate the cost of materials only, which excludes all overhead, labor, profit, permit fees, and rental fees as well as mechanical, electrical, plumbing, and elevator equipment. Furniture is generally be excluded, but it *can* be included if used consistently.

Project teams have the option of calculating the cost of materials in two ways. Project teams can take the total construction costs and multiply by 0.45 to yield a nominal "materials costs only" value, or the team can total the actual materials cost based on a project schedule of values or similar document.

49. *The answer is:* (B) LEED for Schools

LEED for Schools is the only rating system that requires all school projects seeking LEED certification to meet minimum acoustical performance standards. Projects certified under other rating systems may be eligible for an Innovation in Design credit for meeting or exceeding acoustical standards.

50. *The answer is:* (A) the total cost of an asset over its useful or anticipated life

Life cycle cost analysis is fundamental to green building. Project teams performing a life cycle cost analysis are better equipped to see the long-term financial impact of a particular material or system. These annually-recurring costs can add up quickly as many materials or systems are in use for decades.

51. *The answer is:* (A) USGBC membership

The LEED project registration fee is paid upon project registration. The cost of LEED registration varies only depending on whether the registrant is or is not a USGBC member. Registration fees are the same regardless of the LEED rating system used. Once a project team has registered on the GBCI website, members receive information, tools, and communications that will help guide them through the certification process.

52. *The answer is:* (A) building capacity

The full-time equivalent (FTE) occupancy calculation is based on the number of full-time employees, part-time employees, residents, and peak transients (e.g., students, visitors, customers). It is *not* based on the building capacity. The FTE represents a regular building occupant who spends 40 hours per week in the project building.

53. *The answers are:* (D) giving clear guidance to customers
(E) protecting the integrity of the LEED program
(F) reducing challenges that occur during the LEED certification process

All projects seeking LEED certification must adhere to the minimum project requirements (MPRs), which include giving clear guidance to customers, protecting the integrity of the LEED program, and reducing the challenges that occur during the LEED certification process. Certification may be revoked from any LEED project not in compliance with an applicable MPR. USGBC provides more information about MPRs in its "Supplemental Guidance" document.

54. The answer is: (C) zip code

LEED regional priority credits are meant to provide incentive to address geographically specific environmental issues. LEED Online automatically determines a project's regional priority credits based on its zip code. Each regional priority credit is worth 1 point, and each project can earn up to 4 regional priority points. Regional USGBC and local councils have determined six credits in each LEED rating system that are of importance to specific geographic areas.

55. The answer is: (B) participating in a LEED training course

Participation in a LEED training course is an option for compliance with requirements to become a LEED Green Associate, but it is not a requirement for becoming either a LEED AP or a LEED GA. Demonstration of professional experience on at least one LEED project is one of several eligibility requirements for eligibility for LEED AP with specialty exams. Additional requirements include verification of work on a LEED project within the last three years and the participation in the credentialing maintenance program (CMP).

56. The answers are: (B) contractor overhead and profit
(D) labor costs
(E) materials costs

Hard costs are the direct building construction costs. They include direct contractor costs for labor, materials, equipment, and services; costs associated with contractor overhead and profit; and costs of hard assets such as land, inventory, or equipment.

57. The answer is: (C) schematic design, design development, construction documentation, final completion, occupancy certification

USGBC references different phases of the design and construction process throughout the LEED rating systems. The order of the construction process is: schematic design, design development, construction documentation, final completion, and occupancy certification.

58. The answer is: (A) register a project

Project teams interested in seeking LEED certification must first register the project through the Green Building Certification Institute's website, not LEED Online. However, once a project is registered, project teams can use LEED Online to manage project details, complete documentation requirements, upload supporting files, submit applications for review, and earn LEED certification.

59. The answers are: (A) U.S. EPA
(B) NIST

LEED credits are weighted based on the potential environmental impacts and human benefits of implementing the strategies to achieve the credit. To determine which environmental impacts and human benefits are most beneficial, USGBC used the U.S. EPA's Tool for the Reduction and Assessment of Chemical and Other Environmental Impacts (TRACI). TRACI's environmental impact categories were developed to assist with impact evaluation for life-cycle assessment, industrial ecology, process design, and pollution prevention. From these categories, credit weightings were devised under the auspices of the National Institute of Standards and Technology (NIST). NIST was used to compare impact categories and assign a relative weight to each category.

While the American National Standards Institute (ANSI), the BRE Environmental Assessment Method (BREEAM), and the American Environmental Standards Institute (AESI) are all related to LEED in some way, none of their policies are used to determine credit weightings.

60. *The answer is:* (C) testing ventilation filters for removal efficiency

ANSI/ASHRAE 52.2 provides methods for measuring the performance of air cleaners based on their ability to remove particles from the air stream and their resistance to airflow.

61. *The answer is:* (C) the final design review or final construction review

A credit appeal is permitted either after the final design review or after the final construction review. Like a typical credit submittal, the appeal is submitted online. An appeal must be paid for when the project team submits it for review.

62. *The answer is:* (B) 12 months

According to LEED minimum project requirements, in order for a building to be certified as an existing building, it must be occupied continuously for 12 months.

63. *The answer is:* (D) participating in at least 15 hours of continuing education every 2 years

Upon receiving a LEED Green Associate credential, the person must participate in the credentialing maintenance program. The credentialing maintenance program involves completing fifteen continuing education hours biannually. Three of these hours must be completed with an approved LEED focused curriculum.

64. *The answer is:* (D) All rating systems give credits a single, static weight.

All LEED rating systems have a base of 100 points, but there is the opportunity to earn 10 additional points through innovation in design (or operations) and regional priority credits. Therefore, the maximum quantity of points that a project can earn is 110 points. In some rating systems, credits have point values that are not whole numbers (e.g., 0.5). All LEED credits are weighted based on the potential environmental impacts and human benefits.

65. *The answer is:* (B) GBCI

GBCI certifies LEED projects and carries out the LEED credentialing programs. GBCI was established in 2008 as a separate incorporated entity that is supported by USGBC. USGBC handles the development of the LEED rating systems, but it does not certify projects. Prometric is a third-party testing site that administers the professional credentialing exams. The American National Standards Institute (ANSI) sets standards that professional credentials programs must meet.

66. *The answers are:* (A) well water
(B) water from a municipal system

USGBC defines *potable water* as water that meets or exceeds the U.S. EPA's drinking water quality standards and is approved for human consumption by the state or local authorities having jurisdiction. It may be supplied from wells or municipal water systems. Within the Water Efficiency credit category of the the LEED rating systems, USGBC sets requirements for credit earning to encourage project teams to reduce potable water consumption. Many water saving strategies and technologies have minimal upfront cost and can save significant quantities of water.

67. *The answer is:* (A) project site area

The density radius (which is used to determine the boundary of properties that must be included in development density calculations) is calculated using the project site area.

68. *The answer is:* (B) Cost differences between buildings are due to how green a building is.

The Cost of Building Green Revisited found that building cost differences are not based on how green a building is; cost differences are largely based on the purpose of the building. Furthermore, the study found that there are actually no significant cost differences between green and non-green buildings.

69. *The answer is:* (B) percentage of waste diverted from landfills

Within the LEED rating systems' Materials and Resources credits, construction waste management calculations deal with the percentage (by weight or volume) of waste that has been diverted from landfills. Methods of diversion include reusing materials in other applications (both in and out of the construction industry), recycling, composting, and even redirecting the materials back into the manufacturing process.

70. *The answer is:* (D) 50%

For projects seeking LEED certification under the LEED for New Construction rating system, the owner or tenant must occupy at least 50% of the building's leasable floor area. If the owner or tenant is not going to occupy more than 50% of the leasable floor area, then the project will need to certify under the LEED for Core and Shell.

71. *The answer is:* (B) an academic building on an elementary school campus

LEED for Schools is the only rating system that can be used for the construction or major renovation of academic buildings on K-12 school grounds. However, there are other project types that can qualify under either LEED for New Construction or LEED for Schools. Examples of these buildings are nonacademic buildings on a school campus, maintenance facilities, dormitories, postsecondary academic buildings, and prekindergarten buildings. If more than one rating system applies, then the project team must decide which certification to pursue.

72. The answer is: (C) LEED for Existing Buildings: Operations and Maintenance

Any building type has the option to certify with the LEED for Existing Buildings: Operations and Maintenance (EBO&M) rating system. Buildings certified under the LEED for New Construction (NC) rating system are eligible for additional certification only under the LEED EBO&M rating system. Tenant spaces within buildings previously certified under the LEED for Core and Shell (CS) rating system are eligible for LEED for Commercial Interiors certification.

73. The answer is: (C) a product's full environmental cost, from harvesting raw material to final disposal

Life cycle analysis (LCA; sometimes called life cycle assessment) gives project teams a means of considering the relative environmental, social, and health impacts of the available choices when selecting a product, system, process, or service. The life cycle of a material involves extraction, processing, transportation, use, and disposal, all of which can have negative health and environmental consequences. Some of these consequences may include polluting water and air, destroying native habitats, and depleting natural resources.

74. The answer is: (C) a situation in which pursuing credits together rather than separately has an added benefit

Synergies, as related to the LEED rating systems, occur when one green building strategy is applied to a project and more than one LEED credit is accomplished. For example, a project team might install a vegetated roof to help reduce the heat island effect, but it can also result in stormwater runoff reduction and increased energy efficiency.

75. The answer is: (B) occupant productivity

Daylighting, when properly implemented, is the most efficient lighting strategy available. Because natural light produces less heat for the quantity of light delivered, it is a much better source of light than electric lighting. It does, however, need to be controlled, as too much natural light can cause excessive heat gain and poor visual conditions that can lead to eye fatigue. Properly placed windows, exterior and interior solar control devices such as sun shades and light shelves, and an integrated approach to the design of the architecture and electric lighting systems will help to control and make use of the available natural light.

76. The answers are: (B) gravel
(D) open-grid pavers
(E) planting beds

Permeable surfaces, which are also referred to as *porous* or *pervious surfaces*, allow water to infiltrate into the soil to recharge the water table. Examples of permeable surfaces include gravel, open-grid pavers, and planting beds. As this occurs, the pavers and their substrates help to prevent pollutants from entering waterways. *Impermeable* or *impervious surfaces* are solid surfaces that don't allow water to penetrate. The water instead runs off the surface without filtration, pick up sediment and pollutants on its way, and introduces them into the waterways and storm sewer systems. This increased runoff also causes erosion and prevents aquifers and water tables recharge. Furthermore, impervious surfaces such as asphalt often increase a site's heat island effect. Traditional construction and development practices typically use significantly more impermeable surfaces than does green building design and construction.

Practice Exam Part One Solutions

77. *The answers are:* (C) dual-flush toilets
    (D) waterless urinals
    (E) composting toilets

Blackwater is wastewater from toilets and urinals. (Most jurisdictions also classify any water with organic matter—such as that from kitchen sinks—as blackwater.) To reduce the amount of blackwater produced, project teams can install dual-flush toilets, waterless urinals, and/or composting toilets in the building. Wastewater from handsinks, faucets, baths, and showers is considered *graywater*. Therefore, while automated faucet sensors and reduced-flow aerated faucets help reduce water use, water discarded from these fixtures is graywater, not blackwater. Graywater can be reused in flush toilets, mop sinks, and irrigation systems.

78. *The answer is:* (C) reduce pollution from construction activities

While all of the answer options are important and will help to reduce the impact of development and construction processes over the course of a project, the purpose of an erosion and sedimentation control plan is to reduce pollution related to construction activities. The control of soil erosion, waterway sedimentation, and airborne dust generation caused by construction activities is addressed by a prerequisite within the Sustainable Sites credit category.

79. *The answer is:* (A) LEED for Core and Shell

Precertification is available for project teams pursuing certification under LEED for Core and Shell. It is not required for all LEED for Core and Shell projects and is pursued at the discretion of the owner or project team.

80. *The answer is:* (B) LEED Green Associate

Upon passing the LEED Green Associate exam, the designation of LEED Green Associate is earned. The LEED Green Associate title and logo may be used in professional signatures and on business cards.

81. *The answer is:* (B) 40 points

The fewest points a project can earn and still be considered a LEED Certified Building is 40 points. The certification levels are as follows: Certified (40-49 points), Silver (50-59 points), Gold (60-79 points), and Platinum (80+ points). These are the point thresholds for all LEED 2009 rating systems.

82. *The answer is:* (A) LEED for Commercial Interiors projects must have a minimum gross floor area of 250 sq ft (23 sq meters).

MPRs ensure that the project is of the type that the LEED rating system was designed to certify. (USGBC and its entities are forthright with the fact that not all buildings are eligible for certification.) There are eight MPRs. The fourth MPR states that all projects must comply with a minimum floor area requirement; specifically, LEED for Commercial Interiors projects must have a minimum gross floor area of 250 sq ft (23 sq meters). The MPR's also specify that the gross floor area of the LEED project building must be no less than 2% of the gross land area within the LEED project boundary and all certified projects must commit to sharing with USGBC and/or GBCI all available actual whole-project energy and water usage data for a period of at least 5 years.

83. *The answers are:* (A) A fee is applied for each CIR submitted.

(E) The project team must submit a separate CIR for each LEED requirement in question.

All credit interpretation requests (CIRs) must be submitted using LEED Online, not in the format of a letter. They must contain only the inquiry and pertinent background information necessary for a ruling. Rulings on CIRs do not guarantee credit achievement. The project team must still demonstrate and document satisfaction of all applicable credit requirements.

Rulings are project-specific (not precedent setting). Any project team currently pursuing LEED certification cannot refer to or rely upon previous rulings on credit interpretation requests.

84. *The answers are:* (C) Green Rater

(F) LEED for Homes provider

The LEED for Homes provider or Green Rater can assist with the home's preliminary rating. This rating, which determines the appropriate green measures and strategies for the home, is based on a review of the home design and provides the initial score in the LEED for Homes rating system.

85. *The answers are:* (B) energy efficiency

(E) greenhouse gas emission reduction

USGBC has established that energy efficiency and carbon dioxide emission reductions have the greatest potential to positively impact the environment. As such, the weight of a credit is based on reduction of energy consumption and greenhouse gas emissions.

86. *The answers are:* (A) names of project team members involved

(C) description of unique project circumstances related to the credit

Each credit's submittal template has standard required information, including the names of project team members involved and descriptions of unique circumstances. Other required information includes when and where the activities occurred, what strategies were employed, and how these strategies meet the intention or requirements of the submittal path. Including project costs is optional.

87. *The answer is:* (C) final certification review acceptance

Following acceptance of the final certification review, a LEED project may display marketing materials, including the LEED logo plaque, that call attention to the certification. These materials should not be displayed before the final certification review acceptance.

88. *The answers are:* (C) habitat protection or restoration

(D) stormwater design

The civil engineer plays an important role in the documentation of several LEED credits, including those related to site development (including habitat protection and restoration) and stormwater design. Water use reduction may be related to stormwater management through the collection, storage, and use of rainwater, but the water use reduction credits are generally documented by the mechanical engineer.

Practice Exam Part One Solutions

89. *The answers are:* **(B)** equipment costs
(C) land costs

Soft costs are expenses that are not directly associated with the actual construction of the building. They generally include architectural design costs, engineering fees, legal fees, permit fees, financing fees, operating costs after the project is completed, leasing and real estate profits, and advertising and promotion costs. Soft costs do not include equipment or land costs.

90. *The answer is:* **(A)** The lighting power density will decrease and the cooling load will decrease.

The lighting power density (LPD) is a measurement of how much power is used to illuminate an area. The LPD is based on the connected load of electric lighting in an area. When the LPD is reduced, the electric lighting system generates less heat and the cooling load decreases. This design integration helps reduce the energy consumption in the building. In addition, reduced LPD can contribute toward reduced need for mechanical equipment, and thus reduced initial project costs.

91. *The answers are:* **(A)** thermal gain
(B) glare
(D) contrast

Natural light, when used properly, is the most efficient light source available. However, when not controlled properly, natural light can result in increased thermal gain, which is often not desirable, especially in a warm climate. Furthermore, without proper solar control, excessive thermal gain can cause the HVAC system to use more energy. Glazing with a higher solar heat gain coefficient and *U*-factor helps to limit unwanted solar heat gain. Improperly applied daylighting techniques can also cause eye fatigue by increasing glare and contrast. Exterior sun shades, interior light shelves, tinted and fritted glass, along with proper color selection of materials and paints, will all help minimize eye fatigue.

92. *The answer is:* **(A)** They convert hot water or steam into electrical power.

USGBC considers geothermal energy systems that convert hot water or steam into electrical power to be a renewable resource. Currently, the most common means of capturing energy from geothermal sources is to tap into naturally occurring *hydrothermal convection* systems where cooler water seeps into Earth's crust, is heated up, and then rises to the surface. When heated water is forced to the surface, the steam is captured and used to drive electric generators. Geothermal power plants drill their own holes into the rock to more effectively capture the steam.

Renewable energy systems eligible for credit under the LEED rating systems include technologies to capture solar, wind, geothermal, water, or bio-based energy that provide electric power or directly offset space heating, cooling, or water heating energy consumption.

93. *The answer is:* (C) native plants with moisture sensors

Variations in climate conditions have a considerable affect on landscaping and water use. Using native plants (those adapted to the region's climate conditions) is always recommended. Drought-tolerant plants and xeriscaping are recommended for arid (hot and dry) climates. Hardy, resilient plants such as woody plants and evergreens, which can survive extreme low temperatures, are recommended for cold climates. A temperate climate does not have temperature or moisture extremes, so having native plants and installing moisture sensors, which inhibit unnecessary watering, is recommended for temperate climates.

94. *The answer is:* (C) temperate

Passive (natural) ventilation regulates temperature and humidity by integrating non-mechanical systems, such as thermal, wind, or diffusion, to improve indoor air quality and occupant comfort in a building. Buildings in temperate climates (climates without extremes) can more successfully use natural ventilation and passive conditioning than buildings in colder, hotter, or more-humid climates. These strategies are less expensive than mechanical conditioning and provide occupants with individual control over thermal comfort through operable windows and doors.

95. *The answers are:* (C) minimizing energy use
(D) minimizing occupant exposure to chemical pollutants

Prohibiting smoking indoors reduces the need for mitigating secondhand smoke and smoking related contaminants, thus reducing indoor chemicals and pollutants that occupants are exposed to. It can also help help minimize energy use due to the reduced need for energy-using systems to filter the environmental tobacco smoke.

96. *The answers are:* (A) building floor area
(B) property area

The calculation methodology in the LEED reference guides for determining the development density and community connectivity requires the project team to determine the total area of the project site and the total area of the building in order to quantify the development density of the project, as well as the surrounding area.

97. *The answer is:* (A) Conduct a Phase II Environmental Site Assessment.

If after a Phase I Environmental Site Assessment Report (completed according to ASTM E1527-05) a site is suspected of contamination, then a Phase II Environmental Site Assessment (as described by ASTM E1903-97) must be conducted if the project wishes to seek LEED certification. The substances most commonly tested for are petroleum hydrocarbons, heavy metals, pesticides, solvents, asbestos, and mold. If a site is found to be otherwise contaminated following the Phase II Environmental Site Assessment, it must be remediated to meet local, state, or federal enterprise resource planning (ERP) region residential (unrestricted) standards, whichever is more stringent. Schools that are contaminated by past use as a landfill are ineligible for LEED certification.

Practice Exam Part One Solutions

98. *The answer is:* (D) The project team must have at least one LEED AP as a principal member.

Project teams can earn one point for having a LEED Accredited Professional (AP) as a principal member of the design team under the LEED rating system. This person must have successfully passed a LEED AP credentialing exam. Only one point will be awarded regardless of how many LEED APs are on the design team.

99. *The answer is:* (A) 2003 U.S. EPA Construction General Permit

The 2003 U.S. EPA Construction General Permit (CGP) is the referenced standard in the LEED BD&C rating systems SS Prerequisite 1, Construction Activity Pollution Prevention. The CGP provides provisions for construction teams to comply with the stormwater regulations set defined by the National Pollution Discharge Elimination System (NPDES). A link to the CGP is available at www.ppi2pass.com/LEEDresources.

While the CGP itself applies only to sites one acre or more in size, LEED projects of all sizes seeking certification under the BD&C rating systems must comply with either its requirements, or those of a more stringent local standard. Under this prerequisite, project teams can use both structural and stabilization control methods to minimize the negative impacts of erosion on water and air quality.

100. *The answers are:* (B) construction waste management
(E) low-emitting materials use

The general contractor plays an important role in the implementation and documentation of several credits in the LEED rating systems. Generally the contractor is involved in implementing or assisting in implementation of credits that relate to construction waste management and low-emitting materials (such as adhesives and sealants, paints and coatings, carpets, and composite wood). The contractor may also contribute to construction indoor air quality (IAQ) management plans, both during and after construction, as well as to credits that deal with materials and resources used to construct the building, such as materials reuse, recycled content, and regional materials.

Practice Exam Part Two Solutions

Solutions begin on the page that follows.

LEED BD&C Practice Exam — Answer Key

#	Answer	#	Answer	#	Answer
1.	A	35.	C	68.	B
2.	B, C, E, F	36.	B	69.	D, E
3.	B, E	37.	B	70.	D
4.	B	38.	B	71.	C
5.	A	39.	C, D	72.	C
6.	D, E	40.	A, B, E, F	73.	A, E, F
7.	A	41.	D	74.	A, C
8.	A, E, F	42.	B	75.	C, D, E
9.	C, D, E, F	43.	D	76.	A
10.	C	44.	B	77.	A
11.	A, D	45.	C	78.	C
12.	B	46.	C	79.	B, D
13.	D	47.	C	80.	A, B, C, D, E
14.	A, C	48.	A	81.	C
15.	A	49.	A	82.	B
16.	C	50.	B, D	83.	B
17.	A	51.	B, C	84.	B, E
18.	A, B, D, E	52.	B	85.	C
19.	B	53.	A, F, G	86.	A
20.	B, D	54.	D, F	87.	D
21.	C	55.	B, C, D, E	88.	A
22.	D	56.	A, D, E	89.	A
23.	B	57.	B	90.	D
24.	B	58.	D	91.	B
25.	A, C, E, F	59.	D, E	92.	A
26.	B, E	60.	D	93.	B
27.	A	61.	D	94.	A, B
28.	C	62.	B, C	95.	D
29.	B, D	63.	C	96.	B
30.	B	64.	C	97.	C, E, F
31.	A, B, D	65.	B	98.	A, C, F
32.	A	66.	B	99.	A, B, C, E
33.	B	67.	D	100.	C
34.	B, C, E				

Practice Exam Part Two Solutions

1. *The answer is:* (A) EA Credit 1, Optimize Energy Performance

ASHRAE 55 requires that mechanically heated and cooled occupied spaces meet specific temperature and humidity conditions. Conditioning the air provides a more comfortable indoor environment, which increases productivity, but it also increases energy consumption. USGBC considers this relationship to be a *tradeoff*, or a credit relationship in which one green building strategy hinders another.

2. *The answers are:* (B) SS Credit 5.2, Site Development: Maximize Open Space
   (C) EA Credit 1, Optimize Energy Performance
   (E) MR Credit 6, Rapidly Renewable Materials
   (F) ID Credit 1, Innovation in Design

Materials harvested within a life cycle of ten-years or less, such as bamboo and cork, are considered rapidly renewable resources. Using these resources contributes toward meeting the requirements of MR Credit 6, Rapidly Renewable Materials. Using a high-efficiency wall system (such as ICF) provides an opportunity to earn points in EA Credit 1, Optimize Energy Performance. Providing education and interpretive signage contributes toward earning ID Credit 1, Innovation in Design. Minimizing the building footprint and preserving habitat, as is the case in this scenario, can help meet the requirements of SS Credit 5.2, Site Development: Maximize Open Space.

3. *The answers are:* (B) SS Credit 6, Stormwater Design
   (E) MR Credit 4, Recycled Content

Using porous pavement will improve groundwater recharge, reduce stormwater runoff, and remove total suspended solids (TSS), which all contribute toward earning SS Credit 6, Stormwater Design. Since the project's porous pavement is made from pre-consumer recycled materials, it will also contribute toward MR Credit 4, Recycled Content. When one strategy provides multiple benefits, it is considered a synergy. The tradeoff of (or drawback to) the chosen strategy is that black surfaces increase heat islands, counter to the requirements of SS Credit 7, Heat Island Effect.

4. *The answer is:* (B) EA Credit 1, Optimize Energy Performance

Photoelectric daylight sensors help reduce energy consumption and increase the overall efficiency of the lighting system by automatically turning off electric lighting when there is sufficient ambient daylight. Buildings certified under the LEED for New Construction and LEED for Schools rating systems can earn up to 19 points, and Core and Shell buildings can earn up to 21 points in EA Credit 1, Optimize Energy Performance.

5. *The answer is:* (A) option 1's whole building energy simulation

The whole building energy simulation option, which has no restrictions on project type or project size, is the only EA Credit 1 option that will work for this project scenario. Under this option, a project may earn up to 19 points under the LEED for New Construction or Schools rating systems, and up to 21 points under the Core and Shell rating system. Option 2, worth one point, applies only to office occupancies less than 20,000 sq ft (1800 sq meter). Option 3, worth up to three points, applies to projects that are less than 100,000 sq ft (9300 sq meter), but excludes health care, warehouse, and laboratory projects. There is no option 4 for EA Credit 1.

6. *The answers are:* (D) density factor
   (E) species factor

The density factor, species factor, and microclimate factor are needed to calculate the landscape coefficient. The coefficient is used by the civil engineer or landscape architect to calculate the water efficiency of the site's landscaping design. Neither the amount of water treated and conveyed by a public agency specifically for non-potable uses, the area of vegetated roofs, nor the existing site perviousness are needed to determine the reduction in potable water used for irrigation.

7. *The answer is:* (A) calculations

USGBC offers two options for complying with the requirements of EA Credit 5, Measurement and Verification. Both options are based on the IPMVP referenced standard, which describes how to calculate a theoretical energy-use baseline. Option B uses measurement and verification at the system level (also known as the energy conservation measure level). Option D uses measurement and verification at the whole-building level. Option B is suited to smaller buildings, like the community center described in the problem statement.

The energy consumption established by ANSI/ASHRAE/IESNA 90.1 under post-construction operating conditions is used in EA Credit 1. The energy consumption established by the Department of Energy Buildings Consumption Survey is used for EA Credit 6, Green Power.

8. *The answers are:* (A) SS Credit 5.2, Site Development: Maximize Open Space
   (D) WE Credit 2, Innovative Wastewater Technologies
   (E) IEQ Credit 8.1, Daylight and Views: Daylight

Innovation in Design points for exemplary performance are not available for SS Credit 8, Light Pollution Reduction, WE Credit 1, Water-Efficient Landscaping, or IEQ Credit 2, Increased Ventilation.

Projects with no local zoning requirements that provide open space equal to two times the building footprint can earn an exemplary performance point for SS Credit 5.2, Site Development: Maximize Open Space. Projects demonstrating a 100% reduction in potable water use for sewage conveyance can earn an exemplary performance point for WE Credit 2, Innovative Wastewater Technologies. LEED for New Construction and LEED Core and Shell projects that meet daylight requirements for 95% of regularly occupied spaces can earn an exemplary performance point for EQ Credit 8.1, Daylight and Views: Daylight. School projects must achieve daylight for 90% of all classrooms and 95% for other regularly occupied spaces to earn an exemplary performance point for this credit.

9. *The answers are:* (C) lamp lumens for exterior luminaires
   (D) location of property line
   (E) watts per square foot (square meter) of exterior illumination

Compliance with the requirements of SS Credit 8, Light Pollution Reduction, can reduce night sky pollution, light trespass across property lines, and interior light transmittance through exterior windows. The team's lighting designer will need to analyze the site conditions and the performance of the interior and exterior lighting systems to comply with this credit. Lamp lumens for exterior fixtures and the location of the property line are necessary for calculating the amount of light pollution from a project site. Watts per square foot (square meter) of exterior illumination is the measurement of exterior lighting power densities.

The operation sequence of nighttime lighting is needed only for *non*-emergency lighting fixtures. Exterior pavement surfaces may reflect light; however, SS Credit 8 does not address light reflected off pavement surfaces. Watts per square foot of interior illumination does not directly impact light pollution, though it is important for lighting power densities.

10. *The answer is:* (C) a consultant to or employee of the owner

While one of the mechanical engineer's subcontractors could serve as the commissioning agent, LEED encourages teams to choose independent, third party commissioning agents not responsible for the design or construction of the project and hired directly by the owner. Therefore, using a consultant to or employee of the owner would be the better option.

Under EA Prerequisite 1, Fundamental Commissioning of the Building Energy Systems, and for projects over 50,000 sq ft (4600 sq meter), the commissioning agent may be a qualified employee of the owner, an owner's consultant to the project, or an employee of the design and/or construction firms on the project. For projects less than 50,000 sq ft (4600 sq meter) the commissioning agent can also be an individual on the project and/or construction team.

11. *The answers are:* (A) conduct a design review of the owner's project requirements, basis of design, and design documents
    (D) review building operation 10 months after substantial completion

According to the Building Commissioning Association, building commissioning provides documentation of whether or not the building systems function in compliance with criteria set forth in the project documents and satisfy the owner's operational needs. Under EA Credit 3, Enhanced Commissioning, the commissioning agent must conduct one design review of the owner's project requirements, basis of design, and design documents prior to the mid-construction document phase. Secondly, the commissioning agent must review building operation 10 months after substantial completion. The agent must also review the contractor submittals of commissioned systems to confirm compliance with the owner's project requirements and basis of design.

Other members of the design team can develop a systems manual to illustrate optimal operation of commissioned systems and to verify that the requirements for training personnel and building occupants have been met. Developing a plan to account for ongoing building energy consumption is a requirement of EA Credit 5, Measurement and Verification.

12. *The answer is:* **(B)** SS Credit 6, Stormwater Management

A bioswale is a landscape feature designed to slow stormwater flow. The wide, shallow channel traps pollutants and allows water to filter through the soil. Using bioswales can lessen the impact of buildings and developments on local ecosystems.

Stormwater management is a focus of site development in green building. SS Credit 6, Stormwater Design, deals with the quantity and treatment of stormwater leaving the site.

13. *The answer is:* **(D)** 5 points

The project's 40% reduction in interior water use exceeds the minimum water efficiency for both WE Prerequisite 1, Water Use Reduction: 20% Reduction and WE Credit 3, Water Use Reduction. The team would earn 4 points for meeting WE Credit 3 requirements, plus one additional point for education under ID Credit 1, Innovation in Design.

To establish baseline water use for WE Credit 3, the LEED BD&C rating systems use the Energy Policy Act of 1992 and subsequent rulings by the Department of Energy, requirements of the Energy Policy Act of 2005, and the plumbing code requirements of the 2006 editions of the Uniform Plumbing Code and International Plumbing Code.

14. *The answers are:* **(A)** infiltration basin
                                (C) vegetated roof
                                (E) constructed wetland

SS Credit 6.2, Stormwater Design: Quality Control, deals with the treatment of stormwater runoff. Infiltration basins, vegetated roofs, and constructed wastelands are stormwater management techniques that help reduce the amount of total suspended solids (TSS) leaving the site with stormwater runoff. To qualify for this LEED for Schools credit, the project team must have documented proof that the applied stormwater techniques are capable of removing 80% of TSS.

15. *The answer is:* **(A)** flow rate on sprinkler heads

The intent of the Water Efficiency credit category is to minimize interior and exterior potable water use. WE Credit 2, Innovative Wastewater Technologies, provides strategies to reduce demand for potable water and generation of wastewater, or to treat wastewater that is created on-site. Since the flow rate on sprinkler heads does not relate to wastewater, documentation of it is not necessary for WE Credit 2.

16. *The answer is:* **(C)** photovoltaic electricity

Renewable energy strategies generate, or create, energy for on-site use. Strategies for energy efficiency save energy. Using photovoltaic modules is an example of a renewable energy strategy. Photovoltaics actually convert the sun's energy into usable electricity. Options (A) and (B) are examples of strategies that save energy, but they do not generate energy.

A ground source heat pump is not an eligible source of renewable energy for EA Credit 3. It is an earth-coupled HVAC application that uses a vapor-compression system for heat transfer and does not obtain significant quantities of deep-earth heat.

Practice Exam Part Two Solutions

17. *The answer is:* **(A)** EA Credit 1, Optimize Energy Performance, Option 1

EA Credit 1, Optimize Energy Performance, Option 1, Whole Building Energy Simulation, uses the baseline energy performance to demonstrate the percentage improvement in a proposed building's energy performance. In this case, ANSI/ASHRAE/IESNA 90.1 is used to establish the baseline energy use.

The following LEED BD&C credits and prerequisites require establishing a baseline case for comparison.

    WE Prerequisite 1, Water Use Reduction

    WE Credit 1, Water-Efficient Landscaping

    WE Credit 3, Water Use Reduction

    EA Prerequisite 2, Minimum Energy Performance

    EA Credit 1, Optimize Energy Performance

    EA Credit 5, Measurement and Verification

    EA Credit 6, Green Power

    IEQ Credit 3.2, Construction IAQ Management Plan: Before Occupancy, Option 2

18. *The answers are:* **(A)** boiler efficiencies
                          (B) building-related process energy systems and equipment
                          (D) indoor water risers and outdoor irrigation systems
                          (E) lighting systems and controls

International Performance Measurement and Verification Protocol (IPMVP), Volume III: Concepts and Options for Determining Energy Savings in New Construction, provides guidance to project teams for verifying energy performance of their new construction projects. Using either IPMVP Volume III's option B or D, project teams intending to comply with EA Credit 5, Measurement and Verification, must continuously meter boiler efficiencies, building-related process energy systems and equipment, indoor water risers and outdoor irrigation systems, and lighting systems and controls.

USGBC does not require continuous metering of stormwater runoff volumes or daylight factor to lighting systems ratios.

19. *The answer is:* **(B)** 10 years

The LEED BD&C rating systems' MR Credit 6, Rapidly Renewable Materials, encourages projects to incorporate renewable resources in their construction. USGBC considers a resource to be rapidly renewable if it has a lifecycle of 10 years or less.

20. *The answers are:* (B) rough carpentry

(D) wood doors and frames

To meet the requirements of MR Credit 7, Certified Wood, project teams should base calculations on the value of new wood products only, and exclude the value of materials that are salvaged, refurbished, or made with post-consumer recycled wood fiber. These exclusions ensure that project teams seeking points for MR Credit 7 are not penalized for using non-virgin wood.

21. *The answer is:* (C) identifying the commissioning team and its responsibilities

Part of the scope of work for EA Prerequisite 1, Fundamental Commissioning of the Building Energy Systems, and EA Credit 3, Enhanced Commissioning includes: conducting a review of the construction documents prior to midway through the construction document development, and prior to issuing the contract documents for construction; designating a commissioning authority to lead, review, and oversee all commissioning activities; and reviewing the contractor submittals for the commissioned systems. Identification of the commissioning team and its responsibilities is not included.

22. *The answer is:* (D) 75,000 sq ft (7000 sq meters)

The two types of innovation strategies that qualify for ID Credit 1, Innovation in Design, are exemplary performance strategies and innovative performance strategies. The LEED reference guides define exemplary performance strategies as those greatly exceeding the requirements of existing LEED credits. For an exemplary performance point under SS Credit 5.1, Site Development: Protect or Restore Habitat, project teams can protect or restore the greater of 75% of the site area (excluding the building footprint) or 30% of the total site area (including the building footprint) with native or adapted vegetation on previously developed or graded sites. This will exceed the normal credit requirement of 50%.

For this project, the total site area (135,000 sq ft; 12,500 sq meters) minus the building footprint (35,000 sq ft; 3300 sq meters) is 100,000 sq ft (9300 sq meters). 75% of this is 75,000 sq ft (7000 sq meters). 30% of the total site area is 40,500 sq ft (3750 sq meters). The project team must protect or restore a minimum of 75%, or 75,000 sq ft (7000 sq meters) of this space to earn an SS Credit 5.1 exemplary performance point.

23. *The answer is:* (B) ratio of the light transmitted to the total light hitting the surface

Daylighting has proven to be a cost-effective means of providing an interior environment that promotes the well-being of building occupants. When calculating the glazing factor in a building, IEQ Credit 8.1, Daylight and Views: Daylight, considers the value of visible light transmittance (which is the ratio of the light transmitted to the total light hitting the surface).

USGBC defines the glazing factor as a ratio of the interior illuminance at a given point on a given plane to the exterior illuminance during overcast sky conditions. There are many elements influencing this factor, such as window placement, window geometry, floor area, and total window area.

Practice Exam Part Two Solutions

24. *The answer is:* **(B)** FSC-certified wood

USGBC defines rapidly renewable materials as all materials made from any plant or resource that has a harvest life cycle of 10 years or less. Rapidly renewable materials include wool, bamboo, some linoleum components (such as linseed oil, rosin, wood flour, jute, and limestone), and cork. FSC-certified wood is the only option that would rarely be a rapidly renewable material, as it typically has a harvest cycle longer than 10 years. Project teams may earn a point for using FSC-certified wood under the LEED BD&C Certified Wood credit.

25. *The answers are:* **(A)** prime farmland as defined by the USDA in the U.S. Code of Federal Regulations

(C) habitat for any species on federal or state threatened or endangered lists

(E) previously undeveloped land less than 5 feet (1.5 meters) above FEMA's 100-year flood elevation

SS Credit 3, Brownfield Redevelopment, encourages projects to develop on and remediate brownfield sites. Greenfield sites are prohibited under SS Credit 2, Development Density and Community Connectivity. All other choices pertain to SS Credit 1, Site Selection.

26. *The answers are:* **(B)** a high level of system control over thermal comfort

(E) lighting control

IEQ Credits 6.1 and 6.2, Controllability of Systems, encourage individual and/or group control of thermal comfort systems and lighting, which promotes the productivity, comfort, and well-being of building occupants.

IEQ Credit 2 encourages additional outdoor air ventilation to improve indoor air quality, while IEQ Credit 7.2 encourages the assessment of building occupants' thermal comfort over time. IEQ Credit 1 promotes ventilation system monitoring, while IEQ Credit 5 promotes indoor chemical and pollutant source control.

27. *The answer is:* **(A)** structural elements in square feet (square meters)

Documentation of MR Credit 1, Building Reuse, requires calculations based on the area of a building's primary elements. Reused structural and shell elements contribute to MR Credit 1.1, and reused interior nonstructural elements contribute to MR Credit 1.2. The area of reused elements is divided by the total area of elements to obtain the percentage of existing elements reused.

28. The answer is: (C) decreasing the solar heat gain coefficient

The solar heat gain coefficient (SHGC) measures how well a window blocks the sun's heat from entering the building. Decreasing the SHGC means less heat enters the building through the windows. In a building designed for a warm climate, this reduces the cooling load and therefore reduces energy consumption.

Decreasing the albedo of building hardscape materials will contribute to increased heat island effect and thus energy consumption.

USGBC defines the lighting power density (LPD) as the installed lighting power, per unit area. In the United States, this measurement is typically given in watts per square foot. Lower LPDs consume less energy.

Increasing the building's ventilation rates will increase energy consumption, not decrease it.

29. The answers are: (B) fire stopping sealants

(D) aerosol adhesives

IEQ Credit 4.1, Low Emitting Materials: Adhesives and Sealants, requires that all adhesives and sealants used on the interior of the building (such as fire stopping sealants) comply with SCAQMD Rule 1168, and that aerosol adhesives comply with Green Seal GS-36. For IEQ Credit 4, USGBC defines the interior of the building as the area inside of the weather-proofing system. Roofing and stucco adhesives are applied outside of the weather-proofing system.

30. The answer is: (B) $50,000

Compliance with MR Credit 7, Certified Wood, requires that a minimum of 50% of the total value of the project's permanently installed new wood must meet the requirements of a forest certification system recognized by USGBC. Wood that is salvaged or reclaimed need not be included in the total value calculation. Any portion of the wood that qualifies for MR Credit 4, Recycled Content, cannot contribute to earning MR Credit 7.

31. The answers are: (A) within 40 feet (12 meters) of a small manufactured pond used for stormwater retention

(B) within 200 feet (61 meters) of a small constructed wetland created to restore natural habitats

(D) within 50 feet (15 meters) of a small manufactured pond used for fire suppression

The intent of SS Credit 1, Site Selection, is to prevent the development of sites where negative environmental impact would be significant. Areas that are not eligible for SS Credit 1 include prime agricultural farmland, land less than 5 feet (1.5 meters) above a 100-year flood plain, land specifically identified as habitat for species of threatened or endangered species, land within 100 feet (30 meters) of a wetland, or within setback distances, land within 50 feet (15 meters) of a body of water, and public parkland.

Practice Exam Part Two Solutions

32. *The answer is:* **(A)** 30%

The intent of IEQ Credit 2, Increased Ventilation, is to improve indoor air quality and promote occupant comfort, well being, and productivity. To meet the credit's requirements, the project team must increase ventilation rates 30% beyond the standards in ANSI/ASHRAE 62.1. Complying with this credit results in a tradeoff: using additional fan energy and mechanically conditioning air through heating, cooling, humidifying, or dehumidifying will increase the building's energy consumption.

33. *The answer is:* **(B)** ANSI/ASHRAE 62.1

IEQ Prerequisite 1, Minimum Indoor Air Quality Performance, requirements are based on ANSI/ASHRAE 62.1, which provides minimum ventilation rates for both naturally and mechanically ventilated spaces. For naturally ventilated spaces, the standard provides guidance on the size and location of windows. For mechanically ventilated spaces, it explains how to determine the minimum required ventilation rates.

34. *The answers are:* **(B)** floor area
   (C) window area
   (E) visible light transmittance

IEQ Credit 8.1 Option 2 is a prescriptive compliance path for which the LEED BD&C reference guide provides the following equation to determine compliance with the credit requirements. This calculation requires that the window-to-floor ratio (WFR) and the visible light transmittance (T_{vis}) be known.

$$0.150 < (T_{vis})(WFR) < 0.180$$

There are four daylighting strategy options for compliance with IEQ Credit 8.1. For every option, the project team must demonstrate that adequate daylight was provided to a minimum of 75% of the regularly occupied spaces in the LEED for New Construction and LEED Core and Shell. A second point is awarded for LEED Schools where 90% of the regularly occupied spaces receive adequate daylight.

35. *The answer is:* **(C)** dimensions, locations, and angles of exterior shading devices

Sun path diagrams (or charts) determine the location and angle of the sun at different times of day and different times of year for any given location (latitude). The diagrams are used to inform the project team of both direct sun penetrations through glazing and where shading is needed to avoid unwanted solar heat gain. The ideal solution is one that allows enough daylight to enter the space without having unnecessary solar heat gain. Sun path diagrams are typically analyzed at the solstices to find the extreme conditions of the lowest sun angle on December 21 and the highest sun angle on June 21, as well as the equinoxes for an average path location.

36. *The answer is:* **(B)** 3 points

There are two paths to earning points to ID Credit 1. One path (the exemplary performance path) provides the opportunity to earn up to three points for exceeding the performance objectives set for forth within the LEED for New Construction rating system. These points are awarded for achieving double the credit requirements or for achieving the next incremental percentage threshold of an existing LEED credit. The other path provides the opportunity to earn up to five points from Innovation in Design strategies not specifically addressed within the LEED for New Construction rating system. The maximum number of points that can be earned for a combination of innovation and exemplary performance is five.

37. *The answer is:* **(B)** 125,000 sq ft (11,600 sq meters)

The project team must meet the requirements of SS Credit 5.2, Site Development: Maximize Open Space, Option 1. This option stipulates that the vegetated open space within the project's boundary must exceed the local zoning requirement for open space by 25%. The area of vegetated open space required for this credit is calculated as follows.

$$\begin{aligned}
\text{area of vegetated open space required} &= \begin{pmatrix} \text{local zoning requirement} \\ \text{for open space} \end{pmatrix} \begin{pmatrix} \text{credit requirement} \\ \text{for open space} \end{pmatrix} \\
&= \begin{pmatrix} \text{total building} \\ \text{site} - \text{footprint} \\ \text{area} \end{pmatrix} \times 125\% \\
&= (445{,}000 \text{ sq ft} - 45{,}000 \text{ sq ft})(1.25) \\
&= 125{,}000 \text{ sq ft} \quad (11{,}600 \text{ sq meters})
\end{aligned}$$

38. *The answer is:* **(B)** It is considered potable.

Wastewater effluent is typically treated to tertiary standards, but not to the standards of potable water. Effluent can, however, be used for toilet water, irrigation systems, and cooling towers. Living machines or wastewater aquatic systems use localized natural treatment processes, such as microorganisms, fungi, green plants, or enzymes to treat the wastewater generated by building occupants. Using these processes minimizes or eliminates the impact of contaminants on the environment and reduces the building's strain on public infrastructures by reducing the use of energy, chemicals, and water.

Practice Exam Part Two Solutions

39. *The answers are:* (B) a qualified employee of the owner
(C) a qualified consultant to the owner

All project teams seeking LEED certification must meet the requirements of EA Prerequisite 1, Fundamental Commissioning of the Building Energy Systems. For this prerequisite, if the project is less than 50,000 sq ft (4600 sq meters), the commissioning agent can be a qualified and primary member of the design and/or construction team. (USGBC identifies a qualified person as one with commissioning documentation experience for two or more projects of similar size and scope). If the project is 50,000 sq ft (4600 sq meters) or more, as long as the qualified individual is not involved in the primary design and/or construction of the building, the commissioning agent can be an employee of one of the firms working on the project. This may include a qualified employee or consultant to the owner.

Teams also have the opportunity to earn a point for commissioning under EA Credit 3, Enhanced Commissioning, which requires that the commissioning agent must be an independent third party. USGBC encourages project teams to choose independent third party commissioning agents who are hired directly by the owner.

40. *The answers are:* (A) EA Credit 1, Optimize Energy Performance
(B) EA Credit 2, On-site Renewable Energy
(E) EA Credit 5, Measurement and Verification
(F) EA Credit 6, Green Power

The whole building energy simulation model is a three-dimensional computer simulation that provides performance data and estimated energy consumption in buildings. It does this by taking an hourly analysis for 365 days (based on the past 20 years of climatic data) and then estimating the energy consumption of a project.

EA Credit 1, Optimize Energy Performance, Option 1 uses an energy model to quantify the energy efficiency of a building. This value is expressed as the annual energy cost for the building. In EA Credit 2, On-site Renewable Energy, the model estimates the portion of the building's energy that can be provided through on-site renewable sources. This is expressed as a percentage of the annual energy cost. EA Credit 5, Measurement and Verification, uses a model to establish a baseline estimated consumption for a building which is then compared to the actual energy use in the building. EA Credit 6, Green Power, uses a model to determine how much energy a building is going to consume. The owner must then purchase a minimum of 35% of that value from a green-e certified source.

41. *The answer is:* (D) a measure of heat transfer through glazing or some other product

The U-factor is a performance characteristic used to analyze the energy efficiency of windows. It provides a quantitative measurement of the heat transmission through a wall or window system. The lower the U-factor, the greater a window's resistance to heat flow and the better its insulating value. The U-factor is measured in Btu/hr-sq ft-°F (W/sq meter-°C). A window's U-factor is based on the window as an assembly, including the glazing, sill, and frame.

42. *The answer is:* **(B)** developing a project within a density radius of 130,000 sq ft/acre (29,800 sq meter/hectare)

The LEED BD&C reference guide provides objectives for achieving exemplary performance on LEED credit strategies. To achieve an exemplary performance for SS Credit 2, Development Density and Community Connectivity, the development density must exceed 120,000 sq ft/acre (28,000 sq meter/hectare).

A project would need to divert 95% (not 90%) of the construction from the landfill in MR Credit 2, Construction Waste Management. In IEQ Credit 8.1, Daylight and Views, a school project can earn two points by providing daylight in 90% of the regularly occupied areas, but this is not an exemplary performance point. In WE Credit 3, a project can earn four points for a 40% water reduction within the building, but an innovation credit for exemplary performance is not achieved until it realizes a 45% reduction.

43. *The answer is:* **(D)** 13

ANSI/ASHRAE 52.2, which is referenced in IEQ Credit 5, Indoor Chemical Pollutant Source Control, presents methods for testing air cleaners for their ability to remove particulates from the air and their resistance to air flow. The minimum efficiency reporting value (MERV) is based on three composite average particulate size removal efficiency points. In IEQ Credit 5, USGBC requires that in mechanically ventilated buildings, new air filters must have a MERV rating of 13 or better. Prior to occupancy, they must be installed to filter both the return and outside air supply to regularly occupied areas.

44. *The answer is:* **(B)** 21,000 sq ft (1950 sq meters)

MR Credit 1.1, Building Reuse: Maintain Existing Walls, Floors, and Roof, is not applicable to a New Construction or Schools project that includes an addition of more than twice the area of the existing building. The credit is also not applicable to a Core and Shell project that includes an addition of more than six times the area of the existing building. Since this problem pertains to a new building (a LEED NC project), and since the existing warehouse is 10,500 sq ft (975 sq meters), the maximum area of the addition is twice the building's area, or 21,000 sq ft (1950 sq meters).

45. *The answer is:* **(B)** microclimate factor, species factor, and density factor

The landscape coefficient (K_L) represents the volume of water lost via evapotranspiration. It is calculated as the product of the species factor (k_s), the microclimate factor (k_{mc}), and the density factor (k_d).

46. *The answer is:* **(C)** in micrograms per square meter per hour

IEQ Credit 4.3, Low-Emitting Materials: Flooring Systems, uses the Carpet and Rug Institute's Green Label Plus program to evaluate VOC emissions in micrograms per square meter per hour.

Though flooring system emissions are measured in micrograms per square meter per hour (rate per area), the VOC emmissions of furniture is measured in micrograms per cubic meter of air (rate per volume). The VOC content of materials, such as aerosol adhesives, is measured as a percentage by weight, and that of paints is measured in grams per liter minus water and exempt compounds.

Practice Exam Part Two Solutions

47. *The answer is:* (C) secure storage within 200 yards (183 meters) of a building entrance for at least 5% of all peak period building users

In addition to providing secure bicycle storage within 200 yards (183 meters) of building entrance for at least 5% of all peak period building users, the project must provide shower and changing facilities either in the building or within 200 yards (183 meters) of a building entrance for 0.5% of FTE occupants.

In a LEED for New Construction or Core and Shell residential project, covered bicycle storage facilities for 15% or more of building occupants must be provided. In a LEED for Schools project, bike lanes without any barriers must extend to the end of the school property in at least two directions.

48. *The answer is:* (A) 50,000 sq ft (4600 sq meters)

For a project site with no local zoning requirements, the vegetated open space area adjacent to the building must be equal to or greater than the building footprint. Since the building footprint is 50,000 sq ft (4600 meters), a minimum of 50,000 sq ft (4600 sq meters) of vegetated open space must be provided.

49. *The answer is:* (A) credit interpretation ruling

A ruling on a credit interpretation request (CIR) clarifies LEED rating system requirements and submittals. If the project team submits a CIR, the ruling will clarify whether or not the showers in the recreation center can be used to meet LEED credit requirements. Project teams must pay a fee and follow USGBC submittal guidelines for each CIR submitted. Credit interpretation rulings do not guarantee credit award—the project team must still demonstrate and document achievement during the LEED certification application process.

50. *The answers are:* (B) increased initial costs due to increased HVAC capacity
(D) increased operating costs due to higher HVAC costs

IEQ Credit 2, Increased Ventilation, promotes improved indoor air quality and occupant comfort, well being, and productivity. Increasing ventilation would have no effect on solar heat gain. Since the ventilated air would be filtered, it would not contribute to higher outdoor volatile organic compound levels. However, there are potential tradeoffs that should be considered when deciding to pursue this credit. Tradeoffs may include increased HVAC energy costs and decreased energy efficiency from air conditioning, as well as increased upfront costs from needing large capacity equipment.

51. *The answers are:* **(B)** EA Prerequisite 2, Minimum Energy Performance
(C) EA Credit 1, Optimize Energy Performance

Ground source heat pumps (GSHPs) are electrically powered systems that utilize the stored energy of the earth. These systems use the earth's relatively constant temperature to provide heating, cooling, and hot water for homes and commercial buildings. This process creates free hot water in the summer and delivers substantial hot water savings in the winter. Because GSHPs help reduce the amount of energy used in the building, they can contribute toward compliance with the requirements of EA Prerequisite 2 and EA Credit 1. However, they do not actually produce energy, so they will not contribute toward compliance with the requirements of EA Credit 2, On-site Renewable Energy. GSHPs are sometimes confused with geothermal heating systems which convert hot water or steam into electrical power, and which USGBC considers a renewable resource.

52. *The answer is:* **(B)** 60 inches (1.5 meters)

The project team will need to install carbon dioxide sensors in the breathing zone, which is between 3 and 6 feet (0.9 and 1.8 meters) above the floor. 60 inches (1.5 meters) is the only answer choice within the appropriate range. These sensors will monitor carbon dioxide concentrations within all densely occupied spaces—those with a design occupancy of 25 people or more per 1000 sq ft (92.9 sq meters).

53. *The answers are:* **(A)** MR Credit 1, Building Reuse
(G) IEQ Credit 8.2, Daylight and Views: Views

Reusing the doors cannot contribute toward MR Credit 1, Building Reuse, because the area of the addition is greater than the credit limit of two times the area of the existing building. Since the project does not meet the requirements of MR Credit 1, reuse of the doors as light shelves can contribute toward MR Credit 2, Construction Waste Management, as they are diverted from the landfill. They could also contribute toward MR Credit 3, Materials Reuse, as reused material. Materials that meet the requirements of MR Credit 3 include both those previously used in the existing building, or at a different location. However, the reused doors cannot simultaneously meet the requirements of MR Credit 2 and MR Credit 3, Materials Reuse. Reusing the doors can also contribute toward MR Credit 5, Regional Materials, since the materials meet the credit requirement that they must have been salvaged from within 500 miles (800 kilometers) of the site.

Reusing the doors as light shelves can contribute toward EA Credit 1, Optimize Energy Performance. The light shelves will reduce energy consumption by providing a better distribution of daylight and by reducing the need for daytime electric lighting.

If the project team uses Option 2 (a prescriptive approach) from IEQ Credit 8.1, Daylight and Views: Daylight 75% of Spaces, there must be sunlight redirection and/or glare control devices. Light shelves meet this requirement. Furthermore, if the project team uses Option 1 (computer simulation) or Option 3 (measurement), the light shelves will help distribute the daylight further into the building and provide a more uniform illumination level. The reuse of doors as light shelves cannot contribute toward IEQ Credit 8.2, because the intent of this credit is to provide building occupants with a connection to the outdoors through outside views. While the light shelves help distribute and control daylight, they will not create a view to the outdoors.

54. *The answers are:* (D) portable, plug in task lights at all work stations
    (F) bi-level switching in all conference rooms

Compliance with IEQ Credit 6.1, Controllability of Systems: Lighting, requires that the lighting systems provide 90% of the occupants with individual control over their lighting. In addition, it requires that all shared group occupant areas be able to make adjustments to meet the needs of the group. Portable, plug in task lights at all work stations and bi-level switching in all conference rooms would meet the requirements of IEQ Credit 6.1. On/off switching in group spaces will not meet the requirements of this credit. While automated controls such as occupancy sensors and light sensors save energy, they do not provide individuals with control, and therefore do not meet the requirements of this credit.

55. *The answers are:* (B) domestic hot water systems
    (C) HVAC & R systems
    (D) renewable energy systems
    (E) lighting and daylighting systems

The LEED rating systems require that all buildings achieve a minimum level of commissioning through EA Prerequisite 1, Fundamental Commissioning of the Building Energy Systems. The rating systems also offer the opportunity to achieve two points in EA Credit 3, Enhanced Commissioning. To achieve this prerequisite, renewable energy systems, lighting and daylighting systems, domestic hot water systems, and HVAC & R systems must be commissioned.

The other prerequisites and credits that deal with energy systems are: EA Prerequisite 2, Minimum Energy Efficiency; and EA Credit 1, Optimize Energy Performance. Both require meeting and/or exceeding ANSI/ASHRAE/IESNA 90.1. They also deal with the building envelope, the HVAC & R systems, service hot water heating, power distribution systems, lighting, and other equipment.

56. *The answers are:* (A) quantity and type of metering points
    (D) duration and accuracy of metering activities
    (E) availability of existing data collection systems

According to the LEED BD&C reference guide, higher measurement, verification intensity, and rigor will cause higher project costs, both upfront and over time. Factors that affect the cost and accuracy of EA Credit 5, Measurement and Verification, follow.

- level of detail and effort associated with verifying post-construction conditions
- quantity and type of metering points
- duration and accuracy of metering activities
- number and complexity of dependent and independent variables that must be measured on an ongoing basis
- availability of existing data collecting systems
- confidence and precision levels required for analysis

57. *The answer is:* **(B)** renewable energy systems

The LEED BD&C rating systems set minimum energy efficiency requirements through EA Prerequisite 2, Minimum Energy Performance. This prerequisite requires the building envelope, HVAC & R systems and controls, service hot water heating, power, lighting, and other equipment (including permanently wired motors) to meet the standards of ANSI/ASHRAE/IESNA 90.1. Because renewable energy systems are not addressed in this prerequisite, they do not need to meet ANSI/ASHRAE/IESNA 90.1 requirements.

58. *The answer is:* **(D)** end user

Projects can earn MR Credit 7, Certified Wood, by using at least 50% (based on cost) Forest Stewardship Council certified wood-based materials and products. *Chain of custody* is a procedure for tracking all steps and entities involved from the point of extraction to a product's end use. This chain includes all entities involved with the manufacturing, processing, and transportation. It does not include the end user.

59. *The answers are:* **(B)** plywood
        (D) door cores
        (E) particleboard

In IEQ Credit 4.4 Low Emitting Materials: Composite Wood and Agrifiber Products, project teams can earn a point for using wood and agrifiber products (such as plywood, door cores, and particleboard) on the interior of the building that contain no added urea-formaldehyde resins.

The Environmental Health Center, a division of the National Safety Council, describes formaldehyde as a colorless, strong-smelling gas. It is widely used to manufacture building materials and is an effective adhesive for laminating plywood and manufacturing particle board. Formaldehyde is a naturally occurring volatile organic compound (VOC) that is found in small amounts in animals and plants.

60. *The answer is:* **(D)** The project is ineligible.

The team cannot use permanently installed grates, grilles, or slotted systems (the only options for permanently installed entryway systems). Since all LEED for Core and Shell projects must have a permanent entryway system, the project will not be able to meet the requirements of IEQ Credit 5, Indoor Chemical and Pollutant Source Control. However, if this were a LEED for New Construction or LEED for School project, the owner would be able to sign a contract to maintain roll-out walk off mats on a weekly basis by a contracted service organization.

61. *The answer is:* **(D)** mean monthly outdoor temperatures

ASHRAE 55 recognizes that thermal responses in naturally ventilated spaces differ from those in mechanically conditioned spaces. The alternative approach offers a broader range of indoor operative temperatures as a function of mean monthly outdoor temperatures for thermal comfort while assuming light activity by occupants. In addition, the conditions are considered independent of humidity, air speed, and clothing. In general, occupants play a larger role in managing their thermal comfort in naturally ventilated buildings through operable windows, whereas mechanically conditioned spaces are typically programmed to maintain relatively constant thermal conditions throughout the year.

Practice Exam Part Two Solutions

62. *The answers are:* (B) the building's estimated annual electricity consumption
    (C) U.S. DOE's Commercial Building Energy Consumption Survey database

To earn EA Credit 6, Green Power, projects must purchase 35% of electricity through a renewable source. Renewable sources are those that meet the Center for Resource Solutions' Green-e Energy product certification requirements. When an energy model is created for EA Credit 1, Optimize Energy Performance, the data provided from the model forecasts the project's electrical needs. When an energy model is not created, the U.S. Department of Energy's Commercial Building Energy Consumption Survey database is used.

63. *The answer is:* (C) 60%

The building's water use reduction is calculated as follows, where total potable water applied is TPWA and total water applied is TWA.

TPWA = TWA − reuse

 = 43,400 gallons − 3400 gallons

 = 40,000 gallons

$$\text{water use reduction} = \left(1 - \frac{\text{design case TPWA}}{\text{baseline TWA}}\right) \times 100\% = \left(1 - \frac{40{,}000 \text{ gallons}}{100{,}000 \text{ gallons}}\right) \times 100\%$$

 = 60%

TPWA = TWA − reuse = 164,000 liters − 12,900 liters

 = 151,100 liters

$$\text{water use reduction} = \left(1 - \frac{\text{design case TPWA}}{\text{baseline TWA}}\right) \times 100\% = \left(1 - \frac{151{,}100 \text{ liters}}{380{,}000 \text{ liters}}\right) \times 100\%$$

 = 60%

64. *The answer is:* (C) installing thermostats in conference rooms

IEQ Credit 6.2, Controllability of Systems: Thermal Comfort, requires that 50% of the occupants have individual control over their thermal conditions. It also requires that all shared group occupant areas have controls that meet the needs of the group. Thermostats in conference rooms meet this requirement. In addition, IEQ Credit 6.2 requires the HVAC system control at least one of the primary factors of thermal comfort. ASHRAE 55 describes these factors as temperature, radiant temperature, air speed, and humidity.

Operable windows meeting the requirements of ANSI/ASHRAE 62.1 (not ASHRAE 55) can contribute toward earning this credit.

65. The answer is: (B) rate is 0.50 cu ft per second (0.014 cu meters per second); quantity is 6,250 cu ft (180 cu meters)

SS Credit 6.1, Stormwater Design: Quantity Control, Option 1 applies to any site with an existing imperviousness of 50% or less. These are generally undeveloped or lightly developed sites. Option 2 applies to any largely developed site with an existing imperviousness greater than 50%. Based on the project description, Option 2 is most appropriate. In order to meet the requirements of SS Credit 6.2, it is necessary to demonstrate a 25% reduction in the total quantity and rate of the stormwater leaving the site. Option B requires a 25% reduction in both rate and quantity.

66. The answer is: (B) an average of 5 Pa and a minimum of 1 Pa

LEED requires sufficient negative pressure in both of these spaces to prevent or minimize the exposure of building occupants to harmful air containments, such as tobacco smoke, hazardous air particulates, and chemical pollutants. A Pascal (Pa) is a unit of pressure equal to one Newton per square meter. Both IEQ Prerequisite 2, Environmental Tobacco Smoke (ETS) Control and IEQ Credit 5, Indoor Chemical Pollutant Source Control, require an average pressure differential of 5 Pa (0.02 inches of water gauge) and a minimum of 1 Pa (0.004 inches of water gauge) when the doors are closed.

67. The answer is: (D) accommodate at least 90% of the annual rainfall volume

To earn SS Credit 6.2, structural measures for stormwater management and treatment must have the capacity to treat at least 90% of the annual rainfall volume. Structural measures (which include cisterns, manhole treatment devices, ponds, and subsurface stormwater filters) are preferred for urban constrained sites with significant imperviousness because these measures require minimal space. LEED documentation requirements are based on the type of control measure. Choosing structural measures instead of non-structural measures for stormwater management changes the requirements.

68. The answer is: (B) increased contaminants from outside air

To meet the requirements of IEQ Credit 3.2, Construction Indoor Air Quality Management Plan: Before Occupancy, the building must maintain an interior temperature of 60°F (16°C) and 60% relative humidity, causing substantial energy consumption. In addition, earning the credit will add costs for labor and materials, and may potentially add time to the project schedule.

69. The answers are: (C) MR Credit 3, Material Reuse
(E) MR Credit 5, Regional Materials

Because the bridge steel is a salvaged material, the project team can earn a point for MR Credit 3. The bridge steel was also extracted, manufactured, and/or processed within 500 miles of the project, so using it also contributes toward compliance with the requirements of MR Credit 5. Even though USGBC has designated that steel have a recycled content of 25% post-consumer recycled content, reused building materials cannot contribute toward MR Credit 4, Recycled Content.

70. The answer is: (D) $35,000

To earn exemplary performance for MR Credit 6, Rapidly Renewable Materials, 5% of the total materials cost of a project be must invested in rapidly renewable materials. The total value of materials can be provided by the general contractor based on actual costs or as an estimate of 45% of the construction costs. In this case, the materials cost was estimated at $1,000,000. 5% of $1,000,000 is $50,000, so at least $50,000 must be spent on rapidly renewable materials. Since the team has already committed to spending $15,000 on bamboo flooring (which is rapidly renewable), $35,000 remains to be spent. Note that these cost requirements exclude all labor, overhead, profit, rental fees, and so forth.

71. The answer is: (C) annual energy costs

EA Credit 1, Optimize Energy Performance's Option 1 (Whole Building Energy Simulation), uses the performance rating method to calculate the total energy consumed in a building. This method determines the building's annual energy cost. Values for energy consumption by fuel type based on a dollar value are broken out using this method. The energy efficiency is expressed as a percentage comparing annual energy costs to a baseline established by ASHRAE 90.1.

To comply with the requirements of EA Credit 1, the project team must run a minimum of five energy simulations, one for the proposed design and four for baseline designs. The average of the total projected annual energy costs for these four baseline design simulations becomes the baseline, which is used to calculate the percentage improvement in building performance.

72. The answer is: (C) September 21 at 9 am and 3 pm

IEQ Credit 8.1 provides four options to earn one point for daylighting strategies in a building. A second point is awarded for LEED schools where 90% of the regularly occupied spaces have been provided with daylight. For any of the options, a project team must demonstrate that adequate daylight was provided to a minimum of 75% of the regularly occupied spaces. The first option allows the design team to run a daylight simulation model to demonstrate that daylight illumination levels are between 25 and 500 footcandles (270 and 5380 lux). These levels must be simulated and met under clear sky conditions on September 21 (the autumnal equinox) at 9 am and 3 pm. Only areas that meet these light levels count toward the total of daylit spaces. However, designs that incorporate view-preserving automated shades for glare control need to comply with only the 25 footcandle (270 lux) minimum.

73. The answers are: (A) window assemblies
(E) structural supports
(F) nonstructural roofing material

For LEED for New Construction and LEED for Schools, MR Credit 1, Building Reuse, is divided into two areas that address existing buildings. MR Credit 1.1 deals with maintaining existing walls, floors, and roofs, while MR Credit 1.2 deals with the reuse of interior non-structural materials. In both cases (and in LEED Core and Shell), the following materials should be excluded from the calculations: non-structural roofing materials, window assemblies, structural support materials, and envelope materials deemed to be structurally unsound.

74. *The answers are:* (A) 15,000 sq ft (1400 sq meter) retail building
    (C) 45,000 sq ft (4200 sq meter) warehouse

EA Credit 1, Optimize Energy Performance's Option 2 requires project teams to comply with the prescriptive measures described in ASHRAE's *Advanced Energy Design Guide for Small Office Buildings, Small Retail Buildings, Small Warehouse and Self Storage Buildings*, or *K-12 School Buildings*. These compliance paths include recommendations for roofs, walls, floors, slabs, doors, vertical glazing, skylights, interior lighting, ventilation, ducts, energy recovery, and service water heating. The following restrictions apply to Option 2.

- Offices and retail buildings must be less than 20,000 sq ft (1900 sq meters).
- Warehouses and self storage buildings must be less than 50,000 sq ft (4600 sq meters).
- K-12 school buildings must be less than 200,000 sq ft (19,000 sq meters).

75. *The answers are:* (C) mechanical ventilation systems
    (D) natural ventilation systems
    (E) mixed mode ventilation systems

The three basic methods used to ventilate buildings are mechanical (active) ventilation, natural (passive) ventilation, and mixed-mode (combined mechanical and natural) ventilation systems. USGBC references these ventilation systems in IEQ Prerequisite 1, Minimum Indoor Air Quality, and IEQ Credit 2, Increased Ventilation. This prerequisite allows different performance metrics and compliance paths for each system.

76. *The answer is:* (A) MERV 8 filters

ANSI/ASHRAE 52.2, which is referenced in IEQ Credit 5, Indoor Chemical Pollutant Source Control, presents methods for testing air cleaners for their ability to remove particulates from the air and their resistance to air flow. The minimum efficiency reporting value (MERV) is based on three composite average particulate size removal efficiency points. In IEQ Credit 3.1, USGBC requires that if permanently installed air filters are used during construction, filtration media with a MERV 8 rating must be used at each return air grille, and that this media must be changed out prior to occupancy. It is better to avoid using permanently installed air handling units during construction.

77. *The answer is:* (A) whole building energy simulation model

The whole building energy simulation model is a three-dimensional computer simulation that provides performance data and estimates energy consumption in buildings. It does this by conducting an hourly analysis over a 365 day period, based on the past 20 years of climatic data.

In the case of IEQ Credit 2, Increased Ventilation, this tool can be used to estimate additional operational costs incurred due to increased energy consumption resulting from increasing fresh air rates (e.g., from increased fan use and mechanical conditioning). In Arizona's climate, this may be expensive and should be evaluated based on the building's needs.

78. *The answer is:* (C) fraction of solar radiation admitted through a window or skylight

The solar heat gain coefficient is a performance characteristic used to determine the energy efficiency of windows. It provides a quantitative measurement of the solar heat transmitted, absorbed, and released into the building. It is the fraction of solar radiation admitted through a window or skylight and is expressed as a number between 0 and 1. The lower a window's solar heat gain coefficient, the less solar heat it transmits and the greater its shading ability. Unlike the U-factor in windows, which is always based on the window as an assembly, the solar heat gain coefficient is typically a performance characteristic of the glazing itself. It can, however, be used to refer to the entire window assembly.

79. *The answers are:* (B) Provide full cutoff luminaires in the pedestrian areas.
    (D) Reduce the wattage of the lamps in the pedestrian areas.

The first corrective action to reduce the amount of uplight is to install full cutoff luminaires on the pedestrian poles. These luminaires have no candela intensity at angles of 90 degrees or more above the vertical axis (nadir or straight down). Additionally, the candela per 1000 lamp lumens does not exceed 100 lumens per candela (10%) at an angle of 80 degrees from vertical.

The second corrective action is to reduce the size of the lamps on the pedestrian poles. Under SS Credit 8, Light Pollution Reduction, the amount of uplight is based on the total quantity of lumens produced on site. While safety and other design considerations should be evaluated, reducing the lamp wattage on the pedestrian poles will help reduce the total quantity of uplight produced by the exterior lighting systems.

80. *The answers are:* (A) HVAC protection
    (C) pathway interruption
    (E) scheduling

The Sheet Metal and Air Conditioning Contractors' of North America (SMACNA) Guidelines recommend indoor air quality (IAQ) control measures in five areas: HVAC protection, source control, pathway interruption, housekeeping, and scheduling.

81. *The answer is:* (C) occupancy sensors

Corrective actions for thermal environmental problems should be based on ASHRAE 55 and may include, but are not limited to, control adjustments (e.g., temperature set points, schedules, operating modes), diffuser airflow adjustments, and solar control. Occupancy sensors are typically used to turn off mechanical and electrical systems when rooms are vacant. Occupancy sensors are an energy efficiency strategy, not a thermal comfort strategy.

82. *The answer is:* (B) 5 spaces

SS Credit 4.4, Alternative Transportation: Parking Capacity, Option 1 requires a project's parking capacity to meet (but not exceed) minimum local zoning requirements. It must also allocate 5% of the total parking spaces to preferred parking for carpools or vanpools. Since there are 100 parking spaces included in the design of the project, 5 parking spaces must be provided for carpools or vanpools.

83. *The answer is:* **(B)** a site 55 feet (17 meters) from a wetland

SS Credit 1, Site Selection, prohibits projects from developing buildings, hardscapes, roads, or parking areas on portions of a site that are within 100 feet (30 meters) of any wetlands, isolated wetlands, or areas of special concern identified by state or local rule. SS Credit 1 also prohibits buildings with wetland setback distances as prescribed or defined in state or local regulations, rules, or laws. In addition, it prohibits projects from building on undeveloped land that is within 50 feet (15 meters) of a body of water (e.g., seas, lakes, rivers, streams, and tributaries).

84. *The answers are:* **(B)** gymnasium
(E) parking lot

Mixed-use buildings reduce the need to develop land. Schools run on fixed and predictable schedules, providing the opportunity for the community to use facilities when the primary occupants are not present. To comply with the requirements of SS Credit 10, Joint Use of Facilities' Option 1, project teams must collaborate with the decision-making body of the school and make available to the large community at least three of the possible six school spaces. Teams can choose from the following facilities: auditoriums, gymnasiums, cafeterias, classrooms, playfields, and parking lots.

85. *The answer is:* **(C)** Neighbors of the proposed site state the land was previously used as a landfill.

A Phase I Environmental Site Assessment (as described in ASTM E1527-05) is a prerequisite for schools seeking LEED certification. An environmental professional should complete the assessment, but the person does not need to be a member of the design team. Unless the site is on a landfill, the prerequisite will be met if contaminants are remediated to residential levels. Any school site on land previously used as a landfill is not eligible for LEED certification.

86. *The answer is:* **(A)** a dishwasher that uses no more than 1 gallon (3.8 liters) of water per rack

Process water use reduction decreases the load on municipal water supply and treatment facilities. Schools typically have process equipment in addition to commercial fixtures, fittings, and appliances. This additional process equipment must have documented performance showing at least a 20% reduction in water use over industry standards. Garbage disposals and refrigeration equipment with once-through cooling using potable water may not be included in the calculation. In addition to dishwashers, the LEED BD&C reference guide lists required efficiency standards for clothes washers, ice machines, food steamers, and pre-rinse spray valves.

87. *The answer is:* **(D)** installed annual water use rates

Calculations for WE Credit 3, Water Use Reduction, are based on estimated occupant usage. Project teams estimate occupant usage by determining the occupant type (i.e. full-time, part-time, transient, or residential) and the number of occupants. In some cases, it will be in the best interest of the project team to further divide the facility into fixture usage groups. According to USGBC, these groups are subsets of washroom facilities used by different types of occupants. Installed annual water use rates are not needed for WE Credit 3 calculations.

Practice Exam Part Two Solutions

88. *The answer is:* (A) proof of FSC certification for each wood product purchased

Chain of custody (COC) certification is given to companies involved in processes related to the Forest Stewardship Council's (FSC) certified wood. On all invoices with FSC-certified products listed, the vendor must provide the COC certification number. Project teams may buy non-FSC certified wood, however, and therefore proof of FSC certification is not always necessary. Any invoices for wood must include a line-item list of wood products purchased and their corresponding costs.

89. *The answer is:* (A) gymnasium

LEED defines a regularly occupied space in a school building as a space used by occupants for one or more hours per day to perform teaching, learning, and work-related activities. A gymnasium is such a space. Auditoriums, locker rooms, and main lobbies are considered non-regularly occupied spaces. These spaces are for transient occupants and are not regularly occupied for one or more hours per day.

90. *The answer is:* (D) installing storm windows

Background noise is generated from outside and inside sources, including highway, aircraft, and rail noise, as well as HVAC, media equipment, and self-noise. In many cases, windows are the weakest link in wall construction both thermally and acoustically. Strategies that increase window thermal performance, such as storm windows, will increase window acoustical performance. All other listed options actually increase background noise and reverberation.

91. *The answer is:* (B) IEQ Credit 3.1, Construction Indoor Air Quality Management Plan: During Construction

Having a properly functioning HVAC system is key to mold prevention because HVAC ductwork is susceptible to mold growth. Therefore, a LEED for Schools project team intending to comply with the requirements of IEQ Credit 10, Mold Prevention, must meet the requirements of IEQ Credit 3.1, Construction Indoor Air Quality Management Plan: During Construction. Compliance will result in the proper functioning of the HVAC system.

92. *The answer is:* (A) 35 dBA

The required level of indoor noise for IEQ Credit 9, Enhanced Acoustical Performance, is 40 dBA. An indoor noise level of 35 dBA make the team eligible to earn exemplary performance for IEQ Credit 9.

93. *The answer is:* (B) background off gassing

ANSI/ASHRAE 62.1 demonstrates how to determine the amount of outdoor air needed to ventilate people-related source contaminants and area-related source contaminants. Breathing zones affect outdoor air rate. Zones must be considered when calculating ventilation rates for IEQ Prerequisite 1, Minimum Indoor Air Quality Performance, and IEQ Credit 2, Increased Ventilation. According to USGBC, people-related source contaminants of outdoor air affect occupancy density and activity, and area-related source contaminants of outdoor air affect background off gassing from building materials and furniture.

94. *The answers are:* (A) Design charrettes reveal potential community alliances and partnerships.
 (B) Design charrettes reveal synergies and tradeoffs among green building strategies.

A *design charrette* is a collaborative session involving the design team and project stakeholders. Stakeholders include, but are not limited to, ecologists, environmental engineers, developers, civil engineers, architects, government officials, neighbors, building residents, and surrounding businesses. The goal of a charrette is to develop a comprehensive project plan, including potential green building strategies and opportunities for community involvement. A design charrette could be led by a LEED AP; however, it is not required. A design charrette is not required for LEED certification.

95. *The answer is:* (D) ASHRAE 55

In IEQ Credit 7, Thermal Comfort, the LEED BD&C rating systems address occupant comfort based on the thermal qualities of the indoor environment. In IEQ Credit 7.1, the intent is to provide a system based on industry standards that provides a comfortable indoor environment. In IEQ Credit 7.2, the intent is to verify that the building occupants find the thermal comfort design and installation comfortable. If this system is not comfortable to the building occupants, then the credit requires that there be a corrective plan in place to address the problems. The project must meet IEQ Credit 7.1 in order to attempt IEQ Credit 7.2. LEED Core and Shell is not applicable to IEQ Credit 7.2.

The requirements of IEQ Credit 7.2 state that the owners of the building must conduct a survey of building occupants (adults and students Grades 6 and above) 6 to 18 months after occupancy. Corrections and adjustments must be made to the HVAC systems where surveys indicate problems in comfort for more than 20% of the regular building occupants. The owners must use ASHRAE 55 as a reference for environmental variables in potential problem areas.

96. *The answer is:* (B) 15 feet (4.5 meters) beyond surface walkways, patios, surface parking, and utilities less than 12 inches (0.3 meters) in diameter

A project team can protect undeveloped land by minimizing development. Teams must limit site disturbance to 15 feet (4.5 meters) beyond primary roadway curbs and main utility branch trenches. Project teams must also limit site disturbance to 40 feet (12 meters) beyond the building perimeter, 10 feet (3 meters) beyond surface walkways, patios, surface parking, and utilities less than 12 inches (0.3 meters) in diameter, and 25 feet (7.5 meters) beyond constructed areas with permeable surfaces that require additional staging areas.

Practice Exam Part Two Solutions

97. *The answers are:* **(C)** SS Credit 2, Development Density & Community Connectivity
(E) SS Credit 6, Stormwater Management
(F) SS Credit 8, Light Pollution Reduction

While the project site boundary is important when dealing with most Sustainable Sites credits in the LEED BD&C rating systems, there are only a few credits that directly reference the actual boundary. In SS Credit 2, the project boundary is used to calculate the project site area and the development density of the project. SS Credit 6 deals with the total quantity of water that falls on the project site; the project boundary is used to calculate the total quantity of water on the site. In SS Credit 8, the project site boundary is used to calculate the light trespass on the project site.

98. *The answers are:* **(A)** untreated wood waste, including mill residues
(C) agricultural crops or waste
(G) animal waste or other organic waste

EA Credit 2, On-Site Renewable Energy, provides the opportunity to earn points for producing energy and reducing the need for fossil fuels. USGBC has identified the following biofuels as renewable energy: untreated wood waste, including mill residue; agricultural crops or waste; animal waste or other organic waste; and landfill gas.

Fuels that are not eligible under this credit are municipal solid waste; forestry biomass waste other than mill residue; wood coated with paints, plastics or formica; and wood treated for preservation with materials containing halogens, chlorine compounds, halide compounds, chromate copper arsenate, or arsenic. If more than 1% of the wood fuel has been treated with these compounds, the energy system is ineligible.

99. *The answers are:* **(A)** turf-grass fields for physical education
(B) public roads
(C) concession stands used only during school sporting events

When calculating the development density, undeveloped public areas (such as parks and bodies of water), public roads, and public right-of-ways can be excluded from the project site area. Additional exclusions for schools include physical education facilities (such as playing fields) and playgrounds with play equipment. Buildings used only during sporting events (such as concession stands) are also excluded.

100. *The answer is:* **(C)** 80%

The SS Credit 4.1 pedestrian access option gives LEED for Schools project teams the opportunity to earn up to 4 points when 80% of the total student population lives within the walking distance requirements of the credit. Students in grades eight and below must live no more than 0.75 walkable miles (1200 meters) from the school, and students in grades nine and above must live no more than 1.5 walkable miles (2400 meters) from the school. In addition, the project site must have pedestrian access from all residential neighborhoods that house the planned student population.